UnTechnical Writing

How to Write About Technical Subjects and Products So Anyone Can Understand

Also available in *the UnTechnical Press Books for Writers Series*

The User Manual Manual—
How to Research, Write, Test, Edit and Produce a Software Manual

UnTechnical Writing

How to Write About Technical Subjects and Products So Anyone Can Understand

By Michael Bremer

UnTechnical Press, Concord, CA

UnTechnical Writing

How to Write About Technical Subjects and
Products So Anyone Can Understand

By Michael Bremer

This book is dedicated to Linda.

It could not have existed without support and inspiration from
Barbara, Jeff, Wendy and the "other Michael."

Special thanks to Richard, Tom, Debbie, Laura, Bob
and a few others who know who they are.

Published by:

UnTechnical Press

P.O. Box 272896 • Concord, CA 94527

Publisher's Cataloging-in-Publication
(Provided by Quality Books, Inc.)

Bremer, Michael.
    UnTechnical writing : how to write about
technical subjects and products so anyone can
understand / by Michael Bremer. -- 1st ed.
    p. cm. -- (UnTechnical Press books for writers)
    Includes bibliographical references and index.
    LCCN: 99-90429
    ISBN: 0-9669949-0-6

    1. English language--Technical English. 2.
English language--Rhetoric. 3. Technical writing.
I. Title

PE1475.B74 1999   808'.0666

QBI99-650

Contents

Foreword: The Need for Change

Technical writing, as it is generally practiced today, is a well-defined, excellent method of communicating technical information to a technical audience. The problem is that the audience has changed and expanded.

Technology is no longer just for technologists. Yes, there are a lot of engineers, scientists and technicians out there who need to read and learn more, but today, technology is all-pervasive in our society.

Look into any living room. Look at the merchandise at the local mall or department store. Technology's everywhere: camcorders, VCRs, surround-sound audio systems, alarm systems, computers, word processors, spreadsheets—technology for work, technology for entertainment.

The average citizen is becoming more and more dependent on technology. Do most people think of themselves as technologists? No. They are "normal" people who have no interest in engineering degrees, but find that they must deal with techno-stuff on a regular basis. That means absorbing a lot of technical information.

And that's where today's technology writers come in. And that's why, when we write for a nontechnical audience—that is, for most people— we need to take a different approach: the UnTechnical Writing approach.

This book introduces writers, (and aspiring writers, producers, product managers and others who are involved with explaining and selling technology) to the skills, knowledge and attitude that they need to produce clear, understandable explanations of today's technology.

Beyond the actual writing, the UnTechnical approach encourages writers to break out of the Technical Writer box, and to think and learn and do things that, in many companies, writers aren't expected to think, learn or do. Things that overlap with the duties of people in design, marketing, customer service and business. Things that will help writers and companies better understand and communicate with the non-technical world.

"Any sufficiently advanced technology is indistinguishable from magic."
— **Arthur C. Clarke, Technology and the Future**

"Any sufficiently advanced technology is indistinguishable from a rigged demo."
— **James Klass**

Introduction

"I love being a writer. What I can't stand is the paperwork."

— Peter De Vries

"Writing is easy; all you do is sit staring at a blank sheet of paper until the drops of blood form on your forehead."

— Gene Fowler

This chapter, as a well-behaved introduction does, introduces the subject, explains the purpose and structure of the book itself, and sends you off on your merry way, over the pages and through the chapter.

About UnTechnical Writing

Why UnTechnical?

There's a lot of technology in today's world, and there's a lot that has to be written about it.

I've coined the term *UnTechnical writing* to refer to the writing about technology that is intended to be comfortably read and understood by the nontechnical consumer audience, as opposed to the existing term, *technical writing,* which has a pretty bad, even scary reputation outside of technology circles.

We've all—no matter how technical we are—been frustrated and angered by technical documentation at one time or another, and the less technical we are, the more it happens. The goal of UnTechnical writing, its techniques, skill set and attitudes, is to eliminate the frustration, anger and confusion the average person feels when confronted with technology and technical writing. And, along the way, lessen the frustration, anger and confusion that the writer feels when confronted with uncooperative co-workers and corporate cultures that just don't care about the writing.

UnTechnical writing techniques apply to any writing or writer that touches on technology or any complex subjects. This includes everything from user manuals to press releases to help systems to advertising. Many of the techniques apply to any type of writing.

> "For a list of all the ways technology has failed to improve the quality of life, please press three."
>
> **— Alice Kahn**

What If You're Not a "Writer"?

Forget your job title for now.

If any part of your job responsibility is to teach or explain anything to anyone, then, for the part of your job when you prepare and organize information, you're a writer. If you hire, contract or manage people who teach or explain things, then you're a writing manager.

If you are in any way involved with both technology and the mass market, you'll find useful information in this book.

Background—the Origin of UnTechnical Writing

I gained the UnTechnical perspective over a 10-year period working as a writer and manager in the entertainment software industry. Software manuals and their writers have been regularly maligned. Often rightly so. In fact, the first words out of my then future brother-in-law's mouth when we met were, "So, you write computer manuals. I've always wanted to kill one of you guys."

And this sentiment is not a new one:

> "...Then anyone who leaves behind him a written manual, and likewise anyone who receives it, in the belief that such writing will be clear and certain, must be exceedingly simple-minded..."
>
> **— Plato**

OK, back to the story

Always interested in writing of almost any type, I cut my technical writing teeth while working as a technician/designer at a small communication electronics firm. I wrote data sheets, assembly and testing instructions and product-operation handbooks. They were written for our in-house technicians and for our customers—engineers who integrated our products into larger systems. They were short, to-the-point, by-the-book and could cure insomnia at 100 paces. But they gave the needed information to the target audience. The writing was an arduous, boring task, but it was only part of the job, so it was bearable.

After a number of years in that field, I fell into a writing job at Maxis, a start-up software company that made simulation games. In the new job, in spite of the fact that I was writing about games, I had far more technical information to convey. I felt that I needed to take another approach to my writing for three reasons:

1. The audience was different—I was writing for people who may not have engineering degrees.

2. It was an entertainment product—the customers bought the product to relax and have fun, not to use at work, further their careers or pass a college class.

3. These would be fairly long documents, and would take a lot of time and effort. If the process was going to be as boring as my previous technical writing, I'd probably kill myself. So I was determined to enjoy the writing, and write something I'd enjoy reading.

In my first few writing endeavors at this job, I did get a number of complaints. Technical writers would scold me in letters or phone calls about how the rules say this and that, and would tell me how I should have written the manual. But I got a lot more calls and letters from customers who actually *enjoyed* reading the manuals. What with all the clichés about how people never read manuals, and especially not for games, I figured that these fine people were tasteful, brilliant and generally wonderful, but few in number.

After a few years, the company had grown quite a bit. We had shipped millions of games all over the world. I had a staff of three writers working with and for me. Then we got the results of an industry survey. It found that in the software industry as a whole, only 11 percent of the people surveyed read their manuals. That sounded about right. But it also reported that 70% of our company's customers read our manuals. This was delightful, and shocking. I even had to stop billing myself as "the least-read million-selling author in the world."

I figured I was doing something right.

Over the last 10 years, I've done a lot of writing, and hired, trained and edited a number of other writers. Some of the training was by-the-book: here's how we do things around here. Some of it was explaining the local processes. But a lot of it took the form of conversations about our typical customer, the philosophy of writing, communication, working with others in the company, responsibility to company, responsibility to readers, and how to get your job done with the least pain and stress.

This book is a condensed crash course based on those conversations.

About This Book

Who This Book Is For

- Technical writers who want to start writing for a nontechnical audience.

- Writers who already write for a nontechnical audience, and want to do a better job.

- Anyone who is new to writing and wants to write about technology.

- Anyone who has designed a product and needs to write the manual, instruction booklet, help system or even an explanation of the product to get funding.

- Anyone who manages, hires, contracts or subcontracts writers for high-tech writing.

- Anyone who needs to explain anything complex or technical to a nontechnical audience.

- Anyone who wants to understand what it's like to work and survive as a writer in the high-tech world.

What This Book Won't Do[1]

This book won't teach the basics of writing.

This book won't teach you how to type or how to use a word processor.

This book won't teach grammar.

This book won't cover writing for programmers, engineers or scientists, a.k.a., standard technical writing.

What This Book Will Do

This book will try to make clear the special circumstances of and techniques for writing about technology for the consumer audience.

This book will give you an insider's view of what it's like to work as a writer in a high-tech company.

---

[1] *There are a lot of other books that can help you learn all these things. Check the Recommended Reading section.*

"Don't judge a book by its cover, but by its coverage.[2]"

— **Nobody That'll Admit It, But Definitely Not a Graphic Artist**

A Personal Note

Unlike technical writing—but very UnTechnical in nature—this is a personal book. In it I tell of personal experiences, battles won and lost, lessons learned and my own philosophy of writing and the writing life.

There is no absolutely right or wrong way to write. What I state here are the things that I believe, and that have worked for me on one or more occasions. I approached this book as if I were summarizing all the information I wanted to share with a new writer on my staff, and I offer you the same challenge that I offered them: unless you can come up with something better, try it this way. If you can come up with something better, teach me.

Feel free to contact me about this book and about your writing experiences, tips, tricks and methods. Contact information can be found later in this book.

Coming Soon to a Page Near You

The next eight chapters of this book will help you write, work, organize, present, wrangle, philosophize, craft, finish and otherwise deal with the work, people, places and problems related to your writing.

If you have basic writing skills, have a love of things technical, and enjoy explaining things to people, the information in this book will help you produce better work and enjoy your writing. You may even get some fan mail.

The information I present is grouped into these chapters:

- **The UnTechnical Writer** covers writer selection, skills, development and attitude.

- **The Nontechnical Reader** covers ways to identify and think about your target audience.

- **UnTechnical Writing** presents a number of handy hints, tips, tools,

---

[2] *Which brings to mind the time the Supreme Court justice was arrested by mistake while participating in an undercover sting. He was, of course, released after his lawyer reminded the court that you "Don't book a judge by its cover."*

techniques and suggestions that can save you time and pain while making your work more effective.

- **Editing** covers the ways and means of fine-tuning your writing and confirming its accuracy.

- **People and Politics** helps you handle the other human beings that you have to deal with to get your work done.

- **Layout and Graphic Arts** covers the polishing and presentation of the document, as well as dealing with graphic artists.

- **Interface Design** covers the basics of understanding and designing products, hardware and software, so they are better suited to humans—even and especially nontechnical humans.

- **Philosophy** is where I really pontificate on lessons learned over many years of writing.

- **Exhibits** are sample forms, lists and worksheets that may be helpful to you in your writing.

A Warning

A lot of this book encourages you to break out of the Technical Writer box, and think and learn and do things that, in many companies, writers aren't expected to do.

This can be liberating, but it can also cause problems. Take it slowly. You can't declare yourself a new person and expect a company to instantly change their ways of doing and thinking. Change is a slow process.

Don't lose your day job while working towards your goals.

The UnTechnical Writer

"Why do writers write? Because it isn't there."

— Thomas Berger

"I'm writing a book. I've got the page numbers done."

— Steven Wright

This chapter talks about the person who does the writing (whether or not the word writer is in their job title), the skills they need and how to get them.

Roles, Goals and Attitude

Who Should Write About Technology and Technical Products?

The obvious answer is: a writer. But in the real world, who actually does the writing?

The Draftee

Often, product documentation for both hardware and software is written by whoever is around and appears to the powers that be to have the time. If there is no staff or team writer, then this could be the producer, programmer, designer, engineer, artist or almost anyone else.

This can work, as long as the person stuck with the job:

1. communicates well in writing,

2. has and takes the time to do a thorough, well-tested, well-edited job, and

3. relates to the audience.

Relating to the audience is a balance between too much technical detail and too little fundamental information.

Extremely technical experts on the product, programmers, for instance, like to describe the intricate internal details that they love and spend their lives with. This is great if the audience is another programmer or someone whose job it will be to maintain the product, but confusing and frustrating for the typical consumer who buys the VCR, word processor or game.

At the other end of the scale, anyone who has been working on a product for months or years easily forgets what it's like to approach the product as a total unknown. After living with it for so long, the basic aspects of the product seem so obvious that they're hardly worth mentioning. The person buying the product and opening it for the first time would disagree.

It's possible to make both these mistakes at the same time. I've read web pages and press releases announcing new products that list the new features and the patented technology that makes the product so great, but nowhere state what the product does.

"Official" Writers

Sometimes a contract writer is brought in at the last minute and given very little time to learn the product through and through, figure out what the user needs and deserves to know, organize the information, write it, test it, rewrite it, guide it through layout (or do the layout), and do a good job. As frustrating and difficult as this is, it's part of the gig. The contract writer who can come in, grasp the gestalt of the product and get the job done on a tight schedule is the contract writer that gets a lot of jobs.

Under ideal circumstances, the person charged with the writing, whether on staff or contract, will be at least somewhat involved in the project for the last 1/4 of development time, if not all along the way. This gives the writer a good grounding in the product ahead of time, so, in that last rush of development crunch-time, the writer can concentrate on writing and accuracy, and not have to figure out the basics. It will also allow the writer to suggest possible interface and design changes that will not only make the product easier to use, but may also eliminate many pages from the manual, which saves both time and money.

Writer As Communicator

This may seem a sacrilege to many writers, but I believe that the world would be a far better place if there were less technical and UnTechnical writing. Don't worry, there will always be a need for writing and writers, but I hope that a lot of what we do now becomes unnecessary.

If products are really well-designed with intuitive, self-explanatory interfaces, they won't require manuals, or at least not manuals that have extensive sections explaining the basics that should be obvious in a half-decent product.

The classic case that illustrates this point is the door. You don't notice well-designed doors. You reach for the handle or knob or whatever, and your hand knows what to do—push, pull, twist, shout, etc.—and you're through. But a badly designed door will give you the wrong message. You'll pull when you should push, or vice versa. And that's where the unnecessary writing comes in. Some doors actually need an instruction manual. The manual may be only one word—push or pull—but it shouldn't be necessary. If the door hardware is designed well, you won't have to stop and read instructions before using it. The presence of that word on the door advertises the fact that the door hardware wasn't tested with actual door users. And that's typical of the many products, hardware and software, that require long, tedious manuals to make up for the fact that the product wasn't designed well or tested with the intended end user.

And who's going to change the way of the world? You are. Yes, you, the current and future technical and UnTechnical writers of the world. You've got the communications skills and the technical background. All you need is some knowledge, experience, a lot of patience, and the right mindset. The knowledge comes from studying product and interface design. The experience comes from working with others from whom you can learn both the right and wrong ways to do things. The patience comes … if you wait for it. The mindset is knowing that you are more than a button-explainer—you are a communicator.

Even now, if you think of yourself as a communicator who passes on information instead of a writer who puts words on paper, you may be able to insert bits of necessary information into the product itself instead of into the manual.

The UnTechnical Writer's Goal

Your goal as an UnTechnical writer is to communicate. To get information from point A (those that know) to point B (those that need or want to know) as quickly, easily, completely, painlessly and even enjoyably as possible.

That's it.

Don't worry about impressing your professor.

Don't worry about winning awards.

Don't worry about creating art.

Just get that information into the customer's brain, ready to use, right away.

The information you pass on can be in writing or pictures, in manuals, on labels, on reference cards or in the design of the product itself. Don't limit yourself. Communicate.

> **"The virtue of books is to be readable."**
>
> **— Emerson**

Point of View: The Writer's Roles

In fiction, the writer writes to entertain or reveal something about life from any point of view imaginable, from child to dead spirit to deity to alien. In standard technical writing, the writer writes to inform and pass on mastery of a system or subject from the point of view of one technical professional to another.

In UnTechnical writing, the goal is also to inform and pass on mastery of a product, system or subject. There are a number of points of view that work, and some that definitely don't.

When writing for the consumer audience (the whole point of this book), avoid the standard technical-writing point of view. Since the target audience isn't a technical professional, it is easy to make assumptions about their knowledge base and leave out important basic facts, put in too many details or move too fast. You should also avoid writing from the point of view of the Master deeming to pass on knowledge to the unwashed masses (unless you can do it with enough self-deprecating humor).

Choosing your point of view depends on the project, your mood and your personality. Unlike in fiction, where point of view needs consistency and deeply flavors the whole project, in this type of writing, point of view is more of a writer's internal state of mind. It's closer to role-playing than the standard literary point of view, although the role you pick doesn't need to be—and probably shouldn't be—obvious to readers.[3] Point of view is for you; it's a tool for setting and controlling the document's level of technology and rate of delivery.

---

[3] *For the right product and audience, you may want to make your internal role obvious. For instance, present the document as a travel guide to visit the land of [insert your product name here]. This kind of thing works better with younger audiences. If you do try it with adults, don't be too heavy-handed, or make them wade through too much fluff. It gets old fast.*

As you think about how you'll tackle a project, try on these roles for size, and see what fits you and the project:

- **Writer as translator**—you're the link between a world that speaks Technese and people that don't know the lingo. Your job is to translate technical words, concepts and processes into normal language.

- **Writer as host**—you welcome your guests to a new, scary, complicated place. Make them feel at home. Treat them with respect. Talk to them in words that they understand and feel comfortable with. Not only the words, but also the format, feel and media as well.

- **Writer as friend**—you're trying to make your friends feel comfortable while helping them master this new experience.

- **Writer as nerd next door**—you're the guy who knows and loves technology, and you're the one everyone on the block comes to when they have technical questions. You know the deeper inner workings of the technical universe, but know most of your neighbors don't care about that, so you reassure them, then tell them what button to push to get the result they want (tutorial) and give them more information if and when they ask (reference).[4]

- **Writer as teacher**—not the cranky teachers you hated, but your favorite teacher that lit that spark in you and made you want to learn.

- **Writer as intrepid explorer**—you've been down this dangerous road before, and are documenting your discovery process for the readers, allowing them to follow and see the sights, yet avoid the pitfalls. This is especially useful for explanations of very complex systems.

- **Writer as tour guide**—you're guiding a group of tourists through a dangerous, foreign land. You've got to watch out for them, even protect them, while showing them a good time.

---

[4] *All those billion-dollar computer companies should set aside some sort of retirement fund for Nerds Next Door. Judging from the complexity of early computers and the incomprehensibility of many early computer manuals, if it hadn't been for those noble Nerds Next Door simply explaining what button to push to get the result you want, computers would never have caught on.*

If none of these roles works for you, find something that does. The idea here is to get into the right frame of mind to communicate with your audience at the right technical level.

The Writer's Duties

The fiction writer's duty is primarily to self and to art. Beyond that, depending on the writer, there may (or may not) also be a strong sense of duty to the reading audience and the paying publisher. Sometimes these duties come into conflict: write the book you want or that your readers want? Create art or a commercial success?

For both technical and UnTechnical writers, duties can conflict as well. Here are the main duties you'll be expected to fulfill, cross-purposes and all:

- **Duty to self**—your desire to do the best work you can, to write a masterpiece, to create art, and to promote your career.

- **Duty to employer**—your need to satisfy the boss for continued or future employment, and have a positive impact on the bottom line. A positive impact on the bottom line includes minimizing costs to the company by controlling the time and cost of writing, as well as the size and cost of documents and their related shipping costs. It also involves the cost of after-sales support (eliminating as many tech-support calls as possible), and improving the opportunity for future sales of updated versions and other company products. (If a customer buys a product and doesn't understand it or can't use it, they won't buy another from the same company.)

- **Duty to team**—whether you've been on the project since the beginning or just joined weeks before shipping, the design/development team has been working on the project for months or years. They deserve docs that "do right" by the product, show it in a good light, make it more useful, and promote it. (And mention all their names—properly spelled—in the credits!)

- **Duty to product**—often the product itself, whether computer game or VCR or whatever, is such a good piece of work that it deserves to be given a good chance to succeed. And that means it should be well-documented, so it can be used and enjoyed as quickly and easily as possible.

- **Duty to customer**—marketing will identify with and represent the customers up to the point where the product is sold, but it's up to you to be there for them once they get home with it. And beyond their absolute needs, what does the customer deserve? What would you demand, expect, or want if you just brought the product home and opened the box? What would satisfy you as a customer? What would satisfy your mother (assuming your mother isn't a technological wizard)?

You may find yourself split between duties, especially between those to self, employer and customer. When this happens, try to find the common ground between duties.

Usually, you'll find that the most common common grounds are the good of the customer and the bottom line.

Nobody Reads It Anyway. Why Bother Doing a Good Job?

At least once in your writing career (probably much more often), you will either hear or make this statement.

Unfortunately, it's often true. And often deserved. Many people stopped reading manuals because they just didn't help, or weren't written for the right audience or were too boring. Or maybe a lot of people out there just can't read.

Let's face it. To many people technical documents are bad jokes. And, unfortunately, the joke's on the reader.

But it's not hopeless. During most of my time at Maxis, the management was supportive of good documentation, and my group was given a lot of leeway and enough budget to do what we thought was a great job.

Our opinion was supported by the survey mentioned in the Introduction, as well as by hundreds of letters, phone calls, emails and comments on product registration cards from readers all over the world who appreciated the work we put into our documentation. These were people who bought a game, yet were impressed and pleased enough to call or write about the manuals.

What does this mean? It means that at the very least it's worth trying to do a good job. Who knows … somebody may actually read it, so it better be good.[5] It's like wearing clean underwear in case you get into an accident.

So what should we do? We should do the best work we can, and slowly but surely change the way the world looks at technical docs. And donate to literacy campaigns, just in case.

And if all else fails to convince you to do a good job, at least think of yourself. Your name is going on it, and if it's good, it can be a great résumé showpiece.

There is recent evidence that the world may be changing. More and more magazines that rate products, especially computers, software, and audio and video products, are rating the documentation as part of the overall package. Read the magazines in the fields you write about. See if and how they value documentation, and use it as leverage, if possible, to get the support and budget you need to do a good job. There's nothing like a review that slams your—or your competitor's— product because of the docs to get the message through to management. Also, try to find out what the reviewers liked or didn't like in the documents they reviewed.

> **"Good books are the most precious of blessings to a people; bad books are among the worst of curses."**
>
> **— Edwin Percy Whipple**

---

[5] *Actually, my long-term studies have shown that at least one person will read any printed material shipped with a product. And it is a universal truth that that one person will find each and every typo, error and omission and let everyone in the universe, including you, your boss, the chairman of the board of the company you did the project for, the current President of the United States, and your mother, know about them. I suspect that this person who reads everything may actually be one single person, on a mission of some sort. His name is probably Larry, although he uses many aliases. Watch out for him.*

Art and Attitude

Attitude is a very personal thing. I won't tell you how to feel about your work, but I will tell you a bit about how I feel about mine.

First, I believe that there is an art to this kind of writing, but the writing itself isn't necessarily art. If others wish to see it as art and appreciate the finer nuances of meaning, the flow of the language, the sparkling wit and all that other artsy-fartsy stuff, that's fine with me. But as far as I'm concerned, if I want to write "art," I'll write poetry, fiction, screenplays, songs or some other set of words. When I work on an UnTechnical piece, I am concerned only with getting the necessary information into the reader's mind as quickly, easily and enjoyably as possible. (This should be sounding familiar to you by now.) Be proud of your work, but don't let that pride lead you to write more, write less or write more complexly than is in the best interest of the reader.

Art aside, the attitude that works for me as an UnTechnical writer is basically schizophrenic. I split my brain between the combination of product developer, team member, and company employee (or contractor) on one side and advocate for the customer and reader on the other. In this "us and them" world, the writer is both.

Know the Local Attitude About Documentation

As writers, full of the immeasurable importance of our work, we believe that a product's documentation is a vital part of the product. It is actually part of the user interface, and can determine whether the customer has a good or bad experience with the product.

I've worked for and with people and companies that came close to my own way of thinking. And I've worked for and with people and companies with the attitude that docs are a necessary evil, and they should be slapped on as quickly and cheaply as possible. Most companies fall somewhere in between.

Contractor or employee, it is important for you to know where you and your work stand in the local psyche. It will help you to gauge your work, plan your time, estimate hours and pages, and choose your battles wisely (should you decide to choose any).

Understanding how a company pigeonholes writing also tells you about your place in the hierarchy, and your clout. Changing a company's attitude about the importance of good documentation can be done, but it is a slow process. If you wish to stay employed or be rehired on a future contract, don't push too hard.

Why push at all? Some people don't. Some people have to. It's a personal attitude and a personal choice that revolves around the question, "Are you on a mission or is it just a job?"

Be Realistic

So, OK, I've been ranting and raving and preaching about who and what you should be and do and how well you should do it. I've talked about missions and duty.

Do I always meet these standards? Heck no.

Do I try? Whenever possible.

Have I ever failed? You bet.

Am I full of crap? Your call.

Sometimes a job is just a job. Probably more often than not. So, to paraphrase Joseph Campbell, follow your bliss, but don't forget to pay the rent.

Skills and Thrills

UnTechnical Writer Skill Set

Here's a list of skills and knowledge that I used to help me hire new writers and editors for my writing department. I didn't expect anyone (least of all me) to have all these skills, but when I hired new people, I chose people who would fill out our weak spots.

Writing and Editing Knowledge/Ability

- Technical writing experience and ability
- Marketing/advertising writing experience and ability
- Creative/Story writing experience and ability
- Instructional writing experience and ability
- Editing skills and experience

Technical Knowledge

- Familiarity with:
    Windows 3.X, 95/98, NT
    Macintosh
    Novell/other networks
- Programming concepts/basics
- Art concepts/basics
- Design and layout techniques and styles

Educational Knowledge/Experience

- Teaching experience
- Knowledge of Educational Frameworks[6]

---

[6] Many states publish educational frameworks that spell out what students should learn about different subjects at various grade levels. If you are working as a writer on products that will be used for educational purposes, whether manuals or teacher's guides, you'll want to be familiar with and refer to the local frameworks in your writing. At a minimum, you can get by with the California and Texas frameworks. Many other states use these as models.

Software Tool Knowledge

- Microsoft Word[7] (Mac/Windows)
- Other word processors (Mac/Windows)
- FrameMaker
- Online help compilers and help systems (Mac/Windows)
- Presentation software (PowerPoint)
- PageMaker
- Quark XPress
- Pixel-based paint programs (Mac/Windows)
- Vector-based draw programs (Mac/Windows)
- High-end graphics processing programs like Adobe Photoshop (Mac/Windows)
- Spreadsheets
- Project Management software

Personal Attributes/Skills

- Sense of humor
- Self-motivation
- Good people skills
- Good management skills
- Good time-management skills
- Interviewing skills
- The ability to quickly learn new programs and procedures
- The ability to finish projects

Design Skills

- Knowledge of and experience with game or software design
- Knowledge of and experience with interface design

---

[7] *I mentioned this word processor by name not because I wholeheartedly advocate it, but because it was the company standard, and file compatibility required that all writers be familiar with it.*

Not all writing departments are going to care about all of the things on this list, but it won't hurt you or your career to know more than you absolutely need to.

Improving Your Skills

Beyond the obvious ways of improving your skills by learning new programs, reading specific books, working with a good editor and practice, practice, practice, here are a number of other activities that will help you grow as a technology writer.

"All the knowledge in the world is found within you."
— Anthony J. D'Angelo, The College Blue Book

"Never stop learning; knowledge doubles every fourteen months."
— Anthony J. D'Angelo, The College Blue Book

"No wonder I keep gaining weight."
— Michael Bremer

Read and Analyze Technical Writing

In spite of its reputation, there is a lot of good technical writing out there. Read it and learn from it. Read any technical manual or booklet you can get your hands on. Work through the tutorials. Pick a button or feature and see how long it takes you to find the explanation. Make a judgement call on the piece: Is it good? Why or why not?

As you read, ask yourself these questions:

- How is it organized?
- Who are they writing for?
- Who should they be writing for?
- What percentage is tutorial?
- What percentage is reference?
- Does the document serve its purpose? How completely?
- Can you quickly find the information you need?
- Is everything clear?

- Is anything confusing?
- Is everything consistent?
- Is anything missing?
- Is there too much fluff?
- Are the graphics useful?
- Are there enough graphics? Too many?
- How does it look? Is it easy on the eyes?
- What collateral material is there? (Quick-Start guides, readme files, reference cards, etc.)
- What about the finer details? Are there typos? Sloppy language?
- Would you be proud to have written it?
- How would you improve it?

Gather a library of good and bad samples. If you don't have the shelf space or cash to collect them, then borrow them and keep a library of reports in a binder. In each report, put answers to the questions above plus a copy of any part of the book that stands out as truly excellent.[8]

Use this library or set of reports as a resource for style, tone and organization whenever you start a new project.

Read Good Popular Science Books

These days, popular science books are doing well. There are a lot to choose from. And many of them contain wonderful, interesting, clear explanations of complex topics.

Read at least a few of these books, especially those concerning subjects you know nothing (or very little) about. The less you know going in, the better you'll be able to judge how much you learned, and how well new concepts and systems were explained.

The differences between popular science books and product documentation include:

- the audience (generally popular science readers are more interested in science and technology than the average consumer),
- the writer's ultimate goal (provide a good read and entertain rather

---

[8] *Of course, by "copy," I mean copy by hand. I would never, ever suggest photocopying copywritten material.*

than get the information from point A to point B as quickly as possible), and

- the structure (closer to a novel than a reference).

Differences aside, there's much to be learned from this writing. Notice the enthusiasm for the subject that many of these books project. See if you can get a little of that enthusiasm into your writing for the product you're working on (but don't overdo it).

Screenplays and Format

Something to know about screenplays is that they have a very specific format. Type size and spacing and margins are all planned out so that one page of screenplay equals approximately one minute of screen time. Think about how much can happen on the screen during a minute of a movie. With a few exceptions, that fits onto a single page. You have to know what to leave out, what to leave in, and how to make it read well with very few words.

Here's your chance to be petty! You can choose whether to be jealous of screenwriters who don't have to worry about designing a format for every piece of work they do, or be smug and superior because their hands are tied and yours aren't.

Read Screenplays

Have you ever read a good screenplay? Try it sometime.

What's amazing about a good screenplay is how complete it is in spite of what isn't there. Unlike a novel, there's no 10-page passage describing how, as the breeze moves the curtain, the sunbeam coming through the window changes shapes. There's no deep description of the location or clothing or even characters unless they play an important role in the plot. There's nothing but the bones of the story: just the action and dialog and just enough more that makes it a good read.

It's almost skimpy. Just enough to tell the story and give you the feel, the mood. But it's compelling.

No, UnTechnical writing shouldn't read like a screenplay. But there's a lesson in craft to be learned there: cull down your prose to what's important, to only what's necessary. Present steps and actions in a specific sequence, making sure the proper information is there when the reader needs it. And, in spite of the sparseness, keep it entertaining through active language, and subtle humor.

Once hard to find, other than from college libraries, film schools, or just about anyone living in Los Angeles, screenplays are now available in many bookstores and libraries. And you can also get a lot of scripts off the web.

Read some screenplays for movies that you like. Take a screenwriting class if you can, or read one of the seven billion books about it. You'll learn a lot. And you'll have fun rounding out your writing skills in new areas.

Read Good Children's Books

Many great examples of clear, concise, and simple-yet-compelling writing can be found in children's books. In particular, those written for the Juvenile audience (book-publisher talk for 8- to 12-year-olds) are a good resource.

No, your book shouldn't necessarily be written for 8- to 12-year-olds (unless that's your target audience). But reading these books, and seeing that it's possible to tell stories with all their drama, passion and action—in a shorter format, using a simpler vocabulary—can have a positive effect on your writing.

Go back and reread the books you loved as a child. Talk to kids and find out which books they like, and why. Most librarians and bookstore employees can help you pick out some good samples.

> "I wrote a few children's books ... not on purpose."
>
> **— Steven Wright**

> "You know how it is in the kid's book world; it's just bunny eat bunny."
>
> **— Anonymous**

Work on Your Interviewing Skills

Interviewing your team members and other local experts on the project is a valuable way of gathering information for documentation.

Interviewing goes beyond mere conversation in that you prepare yourself with questions and a list of subjects to discuss. You listen carefully, and make sure you understand everything, and keep asking questions until you do. If a new subject that you want to cover is brought up, follow through with it, then return to your original agenda.

Unless you're a born multitasker with great shorthand skills, a small tape recorder can make interviewing much better and much faster. Always test your recorder to make sure it works and that everyone's voice is coming through clearly. If the recorder uses batteries, change them regularly. Running out of a meeting to get new batteries, then asking your subjects to start over can be embarrassing.

Even if you have a tape recorder, bring paper and pencil. You or the interviewee may need to make drawings or sketches to illustrate points, and those don't come through well on tape.

Learn Interface Design

"Damn it, Jim, I'm a writer, not an interface designer!"[9] That's what the good space-faring TV doctor would say in your place. Yet we all know that before the next commercial, he'll have that interface designed and the day saved.

Yes, writing and interface design are two separate tasks, usually with separate job titles, but they're closer than most people think.

The interface—and this goes for hardware and software—is the part of the product that the customer sees, feels and uses to make the product do what they want it to do. It separates and protects them from the complex inner workings, while enabling control over those workings.

Most of UnTechnical writing is explaining how to use the interface to control the product. A manual for a word processor doesn't need to teach how the computer stores the characters in memory or on disk. It does need to teach how to interact with the interface (onscreen buttons, menus and dialog boxes) to make the internal workings of the word processor do what the writer wants. A manual for a television doesn't explain how a television works, it tells the customer how to interact with the interface (the remote) to make the TV display the desired show at the proper volume, brightness, etc.

---

[9] *There may be a law in some states that requires all technically-oriented material with any sign of a sense of humor to have at least one Star Trek reference.*

As a technical or UnTechnical writer, you will study, analyze, dissect, and describe in minute detail many, many user interfaces. You will learn about good and bad interface design—especially bad—even if you don't want to. If you actively seek information about interface-design, you'll get a lot more exposure to the good side of things, and, perhaps, improve your career. If it's known that you have interface design skills, you'll be brought into projects earlier, which not only means more contracting hours, but also gives you a chance to make a positive impact on the product.

Warning! Prepare for a minor rant. It is my hypothesis that fully half of the reason technical writing has such a bad reputation is because the interfaces we write about are bad. Garbage in, garbage out. No matter how good a writer you are, how dedicated and caring you are, how diligently you write, edit and test your work, if the product's interface sucks, your manual will suck. OK, rant's over.

Interface design is a big subject, and the chapter on it later in this book will give only a basic introduction to the parts of interface design that writers really should know.

The Nontechnical Reader

"Knowledge is the antidote to fear ..."

— Emerson

"You know, ... everybody is ignorant, only on different subjects."

— Will Rogers

This chapter discusses the target audience and ways for you to identify, identify with and think about your reader.

Analyzing Readers

Know Your Audience and Write for It

As you start each project, establish in your mind or on paper, as exactly as you can, who the target audience is, so you can talk to them in a way they'll understand.

Find out the answers to these questions:

- **What is the average age and age distribution?**

If the product is tightly targeted to a particular age group, you can tailor your writing to that group. Children, teens, 20-somethings, and 30-somethings up through senior citizens all have particular vocabularies, metaphors and attitudes that they identify with and feel comfortable with. You can also make generalizations about how comfortable different age groups feel with technology.

If the product covers a wide age range, you'll have to stay more generic, and cover all your technical bases, even though the 8- to 12-year-olds may already know it all.

- **What is the gender distribution?**

Most products will be created for men and women. For these, you keep your writing clear and avoid sexist language.

For products that are overwhelmingly or exclusively for one sex, you can customize a little to help the audience relate. If the audience is male, you can exclusively use the pronoun "he." If the audience is female, you can exclusively use the pronoun "she." Sexist language will be covered in detail later in this book.

You can also adjust the style and attitude for either a male or female audience. Men who want to get a better feel for how to write for women should read Deborah Tannen's book, *You Just Don't Understand*. Women who want to get a better feel for how to write for men should read a little Hemingway and a lot of comic books.

- **What is the technical expertise level?**

If your audience is fairly well-versed in the product's technology, you can be a little less detailed in your explanations of how to do things. If the audience is totally new to it, then everything has to be explained step-by-step, taking nothing for granted. Nothing.[10]

- **What is the education level distribution?**

This relates to the age question, but also to levels of higher education. The level of education affects the comfort level for vocabulary, sentence length and grammar "liberties."

As a rule of thumb, aim for an eighth-grade reading level for a "generic" adult audience. Not necessarily because the audience isn't capable of reading at a higher level, but because there's no need for it. Reading level is mostly based on sentence and word length. When you're trying to explain something, using shorter sentences and simpler words whenever possible makes it easier on the learner. They only have to concentrate on the subject matter, and not on deciphering your prose.

Most grammar checkers, including those built into word processors, will analyze for reading level.

- **What percentage are repeat customers?**

If a reasonable percentage of the expected new customers are owners of a previous version or a similar product, you'll want to have a section in the docs that points out the differences/improvements in the new version to help the transition to the new product. Run the product by a test group of previous users without any documentation, and see where they get stuck. Is there a feature they used, but can't find any more? Did names of things change?

---

[10] *I once wrote installation instructions for a computer game that came on two floppy disks. I instructed the customer to put in disk 1 and run the install program, then when the Insert Disk 2 message comes up on the screen, put in disk 2 and hit the Enter key. A month or so after the product shipped, someone in tech support forwarded a call to me. The program wouldn't install. The customer said he followed my directions exactly. He put in the first disk and ran the install program. When the message came up on the screen, he put in the second disk, and hit the Enter key. I didn't know what was wrong until he said, "Both the disks are in there, but nothing's happening." My mistake. I misjudged the technical expertise of my audience. From that day on, I always added an instruction to remove the first disk before putting in the second.*

- **Is the audience from any particular professional field?**

This question has a lot more bearing in technical writing geared toward professionals that it does in UnTechnical writing geared toward consumers, but it can still apply. Different professional fields have their own vocabularies, acronyms and familiarity with the technology. If you regularly write for a particular field, such as education, computer, medical or legal, keep your own glossary or style guide of industry terms. Or buy one.

- **Any other special attributes of the audience?**

They could be parents, people in wheelchairs, movie buffs or gardeners, but each of these will have a specific vocabulary and range of allowable attitudes. If you're writing for a special group like this and aren't personally involved, it's important to get a reality check. Run your writing by someone who is involved in the group. You may learn important things like: parents don't usually refer to their children as *rugrats,* people in wheelchairs are tired of hearing the phrase, "let's get rolling," movie buffs prefer to be called film fans and gardeners' thumbs aren't really green. OK, you're not likely to make these obvious blunders, but you will make some if you don't get that reality check.

On many projects, the audience may be obvious from the product. If it's a shoot-em-up video game, the audience is males, 8 and up. If it's a CD to teach children how to read, the product is targeted at 4- to 6-year-olds, but the documentation's audience is parents and teachers.

When the audience isn't obvious—and even if you think it is—there are a number of sources of information available.

Marketing—depending on the size of the company and the quality of the marketing department, they may be able to tell more than you ever wanted to know about the target customer, or absolutely nothing. In a moderately well-run company, somebody in marketing has probably researched and written a report detailing the projected customer base before the product hit the design stage, and updated it well before advertising deadlines approached. Chances are good that it won't occur to anyone that the writers may need a copy of the report. Ask for it.

Technical Support and Customer Service—marketing has a lot of information based on surveys, focus groups, general market data and projections, but the stalwart troops in technical support and customer service are on the front lines day after day. They know the customers intimately, and can give you a realistic gauge of their technical expertise. And they can tell you which parts of a previous product's docs were wrong or unclear and generated customer calls for help.

If you're a staff writer, see if you can spend some time in customer service and tech support, either manning the phones yourself, or listening in. It can be a very enlightening experience.

Something else you might try is to make a customer definition sheet, and put it up on the wall where you work. This sheet should summarize the answers to the questions at the top of this section, and have pictures of typical customers (real or clipped from magazines). Give the customers names, and try to think of them as real people. Imagine they are reading what you write. Do they understand?

Different Readers Learn in Different Ways

There has been a lot of research recently into how people learn. Do a web search on "Multiple Intelligences" and "Howard Gardner" for a list of websites and related books on the subject. Fascinating stuff.

Knowing that people learn in different ways can help you, as a writer, cover as many bases as you can when presenting your information. The two main ways of learning that we constantly deal with (and can deal with) are visual and tactile.

Visual learners like to pick up the information by reading or watching a demonstration. When you can, use pictures, diagrams, flowcharts, sequence photos and video to help these readers.

Tactile learners need to actually do something before it sinks in. If they just read or watch a demonstration, they don't gain or retain all the information they need. They have to physically do something themselves before they learn it. For these learners, you must supply clear tutorials so the reader can personally perform all the steps in order.

Beyond these two main inroads to the reader's brain, there are always the other senses. In the way that fiction writers make scenes more realistic and vivid by evoking all the senses, so should you—at least as many as make sense—when describing a process.

I doubt that putting a user manual on audio tape will be very useful, but audio as an added attraction as part of a computer help file can be helpful. Onscreen video with clear narration can work well. As for the olfactory sense, what the heck ... if including a scratch 'n sniff card with the product helps, do it.

Different Readers Will Read Your Docs in Different Ways

Unlike with a novel, most people don't read product documentation all the way through from front to back. We like to think they might, since we've presented our subject in an organized and sequential manner. Some will, most won't.

Many things affect the way people read technical docs: technical expertise, previous knowledge of the subject, available time, how often they use the product, and personality. You have to write to communicate with as many potential readers as you can.

Here's a small sampling of reader types, and how to make sure they're covered:

- **Thorough Readers:** those who read everything cover to cover, and work through all the tutorials.

This type of reader may be new to the product or technology and want to read everything to avoid mistakes. Thorough readers may also be very experienced people who want to know everything there is to know about the new product or tool.[11]

For these readers, you need to present the document as a whole, linear unit. The organization must be logical, and must build from beginning to end. The tutorial must be well designed and written, thoroughly tested, and well laid out.

---

[11] *This type of reader is typified by the graphic artist/designer I'm working with on this and other books. Any time he gets a new piece of hardware or software, he reads the manual cover to cover. I've even kidded him about wasting time with all that reading—even for a minor software update. But once we settle down to work, he knows the product so well that the work goes faster, better and smoother.*

- **Thorough Dry-Labbers:** those who read everything and "dry-lab"[12] the tutorial.

This type of reader is similar to the thorough reader, but is familiar enough with the product or type of product that they just read through the tutorial, without actually performing the steps.

For these readers, you need to be sure that your tutorial not only lists the tutorial steps that the reader is intended to perform, but also tells—and shows—the results of each step.

- **Foundation Seekers:** those who only read the introduction section, work through (or dry-lab) the tutorial, and figure out the rest on their own.

For these readers, make sure the introduction section completely describes the product and its capabilities, and that the tutorial covers all the basic knowledge the readers will need to use the product on their own.

- **Self-starters:** those who dig into the product on their own to figure out the basics, and only look at the reference section for specific topics and questions.

For these readers, the table of contents and index need to be very complete. Lots of well-named subheadings help, too.

- **Scanners:** readers who quickly glance through the whole book and only read the headings and picture captions, and anything else that catches their eye.

Give these readers lots of thing to catch their eyes: pictures, headings, drawings, quotes, etc.

- **Avoiders:** people who avoid the manual at all costs.

For these readers, you have to be sure that the one time they resort to opening the manual is a good experience. That means it looks good, it reads well, they can find exactly what they're looking for quickly and easily, and, if all goes really well, you catch their attention so they spend a little extra time paging through the manual, and consciously

---

[12] *Here's a case where I never would have thought to define a term, but because a couple of people (including the editor) who read it didn't know what dry-labbing was, I had to explain it. Dry-labbing is a common practice in biology, chemistry and other classes that require laboratory experiments, where you write up the results of an experiment without actually performing it.*

or not, get a sense of the book and what's in it. If these readers have a good experience they might come back.

Kids as Readers

Obviously, vocabulary is a major factor in writing for kids. The younger the reader, the smaller and simpler the vocabulary. You often have to use new words to explain a new subject, but try to avoid using too many too close together. Books that list common reading vocabularies by grade level are available, and useful. Many grammar checkers will also analyze the grade level of a document.

In general, keep both sentence and word length short. Don't talk down to the reader, but keep it as short and simple as it needs to be.

A note about vocabulary: avoid using words and phrases that parents may not want their children to use. An irate customer is one thing; an irate parent is an angry mob.

On a similar note, it's a good idea to be a little more diligent about grammar when writing for kids.

Another note about vocabulary: slang is a double-edged sword. Using current slang for the proper age group can be very effective, and can help give your writing the edge. But getting the age group wrong or using terms that are no longer in vogue can turn your audience against you. TV commercials and some kids' magazines are a good source for the state of slang words.

Structurally, kids' manuals need more of a tutorial than adults' manuals. An adult generally uses a tutorial to get started, then uses the reference when necessary to move on from there. Some kids may need a little more hand-holding, and therefore more tutorial. But it's more than that. Children will use the product more if you give them a mission, a goal—or many goals—to accomplish.

Yet another vocabulary note: be careful not to make your own work sound boring. Call the tutorial an exploration or mission. Call different steps activities, challenges or goals. Make it fun—and sound fun—every chance you can.

One last (I promise) vocabulary note: watch out for the word "kids." As of the writing of this book, the word "kid" is generally accepted among the young, and is used constantly in advertising and product naming. But that can change. If you've got the word "kid" all over your product, and even in the name, you may suddenly get rejected by kids and have to make some quick changes.

Also, be aware that when it comes to technology, most children, unlike their parents, have no fear. An adult who is unfamiliar with technology needs to be led by the hand to give them the confidence to deal with this newfangled gadget, and to get over their fear of breaking it. Kids will jump in and start pushing buttons to see what happens.

In a sense, this makes it easier for you: the kids can and will figure it out on their own. It also gives you the challenge of making the docs useful enough and fun enough to help the kids through the experimental stage more quickly.

Dealing with Readers

Respect the Reader

There used to be an insult: when somebody made a mistake or asked a "stupid" question, someone else would say, "What, are you new here?" I never liked this insult, mostly because I was always on the receiving end. But also because it made newness and inexperience into offenses. It blamed people for things that weren't their fault.

In theory, the insult was just good-natured humor when it was used on people who weren't new, and who should have known better. But more often that not, I heard it used on people that shouldn't know better. Instead of it being a funny "gotcha" between friends, it created and widened the gulf between us and them, the knows and know-nots, making the person who is already uncomfortable in a new situation even more uncomfortable.

When writing about technology for nontechnical readers, it's easy to get smug and look down on the readers. After all, we're skilled techie nerds[13] and they're not. We know our way around this stuff blindfolded. And what we don't know, we can learn quickly and easily. That's why we do what we do.

Many nontechnical readers, on the other hand, *are* new here, and deserve the time it takes to learn enough about their surroundings to fit in and feel comfortable.

Nobody likes to be talked down to. Nobody likes to be called stupid. Everybody deserves respect. If you don't respect the reader, it can show in your writing, and they won't read it.

Give the reader a break. Even the ones who call tech support because their new TV won't work, because you forgot to tell them to plug it in before turning it on.

---

[13] *Object all you want. You know who you are.*

"No one can write decently who is distrustful of the reader's intelligence, or whose attitude is patronizing."

— **E. B. White**

"People are funny."

— **Art Linkletter, book title**

The Reader as Heckler and Helper

Readers are the reason we write. And the reason we sometimes think of changing careers.

There's nothing more thrilling than getting a letter or email from someone who read what you wrote and liked it. It's like acting for shy people   you get the rush that actors get from being on stage, except you don't actually have to stand up in front of a room full of people.

But then again, everyone's a critic. Especially in books about technical subjects, someone will point out every omission, every mistake, every typo, and everything you do that isn't the way they would have done it if they were a writer. And if they are writers, they can be even more adamant in showing their disapproval or offering helpful suggestions.[14]

And the worst part of it all is that sometimes they're right. Read and listen to any and all comments you can about your work. Don't take every comment and suggestion as gospel, or you'll drown in contradictions (and self-pity). But give every comment a chance. If it makes sense, think about it. If it makes no sense, then forget it.

---

[14] *In 1989 or 1990 I wrote the manual for RoboSport, a multiplayer battle-strategy computer game. I received almost exclusively positive comments on the quality of my work from co-workers and customers. Two articles (one magazine, one newspaper) reviewing the game not only mentioned the manual (a rare thing in reviews in those days), but said how good it was and how it enhanced the playing experience. Sitting around smugly one day, I got a phone call from a technical writer telling me how the manual was no good because all technical manuals must follow the format of "Tell them what you're going to say, then say it, then tell them what you said." The guy actually took the time and money to call me to help me out with this information. I use that approach often (for especially technical, complex subjects, and almost always for tutorials), and it works. But I like to have the flexibility to tailor the writing approach and manual organization to the individual product. Sometimes, especially with an entertainment product, it's better to present the information in an entertaining way than a standardized way.*

A big problem for many technical and UnTechnical writers is that they don't get any feedback from customers.

Try to make feedback easy. Leave room for comments on the registration card. Put an email address for documentation comments into the docs. Ask technical support and customer service to forward you any letters or email they get that mentions the docs, either positively or negatively. Find out if marketing can do, or has done any customer surveys or focus groups on documentation. See if customer service can do some quickie surveys, asking call-in customers a quick question or two (that you supply) about the docs.

You'll learn a lot about your writing and your customers. You'll find out that some people missed your point entirely. Try to find out why, and how you can avoid the confusion next time. You may even want to write or call some of these people to find out why you didn't get through.

You'll get piddling little typos pointed out. This is good. You can correct them in the next edition or print run. And, even better, you can use this as ammunition against your editor—he misses a typo, he buys the next beer or latte.

And in this spirit, I invite and encourage all you writers who read this book to send me any problems, typos, mistakes, omissions and opinions you have on my work. You can send compliments and praise as well, if it strikes you. Ways to reach me can be found at the back of this book.

> "The Stones the Critics hurl with Harsh Intent
> A Man may use to build his Monument."
>
> **— Arthur Guiterman**

Customers and "the Edge"

I firmly believe that writing about technology for the nontechnical consumer audience needs an edge. Something to make them notice, chuckle, feel comfortable and KEEP ON READING.

In writing for highly technical people, this is less important, possibly even bad. But the non-techie needs and deserves to be entertained. Putting an edge on your work through humor, wisecracks, enthusiasm, or even just being a little different or interesting can get you noticed. Sometimes not in a good way.

Here's how I look at it....

About 10% of the people out there like (or are accepting and satisfied with) everything you do, no matter what. It can be great or horrible or boring or exciting. It almost doesn't matter. These people are just fine with it. Barring total rudeness, they're on your side, so you don't really have to worry about them.

About 10% of the people out there won't like what you do no matter what. It's either too long or too short (or both), too funny or not funny enough (or both). These people will never be satisfied with what you do. Never. You can't worry about these people, because it won't do any good.

Another 10% are looking for a reason to like your work. Almost anything that stands out and gives them a chuckle or a surprise is good.

Another 10% are looking for a reason not to like your work. Almost anything that stands out and gives them a chuckle or a surprise is bad. This is your (at least my) target sacrifice group. (No, I don't mean sacrifice *them,* just their goodwill.) If you don't put enough of an edge into your work to get a few cranky comments out of this group, then you don't have enough edge to get that other 60–70% excited.

The final 60% aren't looking for anything. They're there. They need you, but don't expect much. If you can show them a good time, wake them up, and still get the information across, you've succeeded. If you can make them feel comfortable or good, and demystify or defang this little bit of technology that has intruded into their lives, you've succeeded. Most of the time, these people won't notice much about your writing. Not that they're bad people, just that they don't expect much from technical writing in general. If you can give your work an edge that gets noticed, and you deliver information in a fun way, you've won.

This group of people is your "win" target.

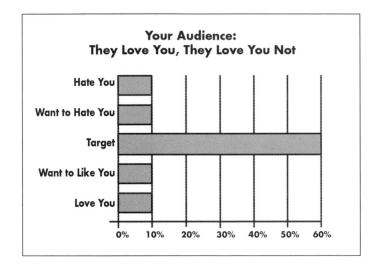

In summary, you can't please all the people, ever. But you can please most of them. Don't feel bad about a few negative comments. Take them seriously, and see what you can learn from them. But if you don't get anyone to complain, then you probably haven't excited anyone either.

Of course, if you get a lot of complaints, then take them very seriously and use them as a guide to your next revision or project.

> "No man can read with profit that which he cannot learn to read with pleasure."
>
> — **Noah Porter**

UnTechnical Writing

"If you would not be forgotten, as soon as you are dead and rotten, either write things worth reading, or do things worth the writing."

— Benjamin Franklin

"I get up in the morning, torture a typewriter until it screams, then stop."

— Clarence Budington Kelland

This chapter covers the actual writing, and touches on working as a writer in the high-tech world.

The Basics

What Is Writing?

Writing, for the purposes of this book, is the act of researching, organizing and presenting information.

What that information is, how it's presented, and whether the research is investigating a technical product or searching your soul depends on what you're writing.

> "Books are the compasses and telescopes and sextants and charts which other men have prepared to help us navigate the dangerous seas of human life."
>
> **— Jesse Lee Bennett**

Learn and Use the Basics of Writing and Technical Writing

All the basic writing skills and techniques that apply to any type of writing apply to UnTechnical writing. It's all communication. There are many good books and classes today that do a great job at teaching writing, including "standard" technical writing. Take advantage of them. Learn from them. You have to know the rules before you can break them well.

Of course, many of the rules will always apply, for example:

- Use of active voice
- Consistent phrasing
- Consistent tense, most often present
- Proper use of words like *then* and *than, affect* and *effect, its* and *it's, their, they're* and *there,* and all those other words that you always find ill-used in email
- Proper punctuation
- Accurate spelling

Once you get beyond the true, invariable, unbreakable basics, you have a lot of leeway. But you have to know the basics. Or at least your editor does.[15]

Since there are so many books and classes to cover them, those basics are not in this book.

Tools of the Trade

As a writer in the technical world, your word processor is your best friend and most indispensable tool. Know its quirks and strengths. Take some time to explore its features.

Here are a few word processor features every writer should know about:

- **Spelling Checkers**

Spelling checkers are wonderful, but they're not a replacement for looking over your own work, much less a reason to bypass an editor. Spelling checkers don't correct instances when you've used the wrong word as long as the wrong word is spelled correctly. And be darn careful about adding your own words to the custom dictionary— once you add in a typo or misspelled word by mistake, your spell checker will never catch that mistake again.

UnTechnical Writing is striking a balance between these two quotes:

"To make communication more meaningful, to make possible wiser judgments on modern complex problems, the ivory tower of Babel must come down."

— Milton S. Eisenhower

Dave Barry, "Tips for Writer's"
Dear Mister Language Person:
What is the purpose of the apostrophe?
Answer: The apostrophe is used mainly in hand-lettered small business signs to alert the reader that an "S" is coming up at the end of a word, as in: WE DO NOT EXCEPT PERSONAL CHECK'S, or: NOT RESPONSIBLE FOR ANY ITEM'S. Another important grammar concept to bear in mind when creating hand-lettered small-business signs is that you should put quotation marks around random words for decoration, as in "TRY" OUR HOT DOG'S, or even TRY "OUR" HOT DOG'S.

Oh, and my editor asked me to add that editors really like it if you use the spelling checker before you give them your manuscript.

---

[15] In my youth I remember sitting with the family and watching Jack Benny play violin on TV. It was hilarious. I said something to the effect of, "He can't play violin at all." My wise mother said, "He's actually very good. You have to be good to play that bad and still be funny." This is somehow applicable to knowing the basics of writing before you can abuse them to good effect.

- **Grammar Checkers**

Grammar checkers are as close to controversial as you can get with a word processor. Some people love them, others hate them. Personally, I find grammar checkers annoying. They always seem to flag typical "problem words" even when they're used correctly. I have friends that say that the newer ones are useful if you take the time to set them up, but I just haven't gotten around to it yet. However, I do use the grammar checker's readability and grade-level statistics.

- **Outlining Mode**

If your word processor has an outlining mode, try it out. Outlining is very useful for organizing and navigating long documents.[16]

- **Tables**

Just about every word processor made in the last 5 to 10 years can make tables. If you're writing in the technical world and your word processor doesn't do tables, it's time to upgrade. But be careful: tables can be lost when exporting and importing files from program to program. Some word processors drop the tables when exporting to RTF (Rich Text Format, a standard interchange format that almost all word processors and page-layout programs can read and write), and some layout programs may ignore or butcher them.

- **Revision Tracking**

Revision tracking is a way of life for some writers and a lot of lawyers. Others never use it.

It can be used with email to solicit comments and corrections to a document from multiple people without printing copies and routing. Maybe it's the way my word processor handles revisions, but I've found that it's more hassle than it's worth on long documents. I'll keep a soft copy and the marked-up edited printouts of each major draft so I can revert if necessary.

- **Macros**

Macros, both those that run processes and those that add boilerplate text can be amazing time-savers. No point in doing extra, repetitive, boring work. That's what computers are for.

---

[16] *Personally, I use the outline view 60–80% of the time I'm writing a technical document. (Less for a short story or screenplay, but still sometimes.)*

- **Templates**

Templates are also time-savers, and can help a group keep a consistent look and feel.

Besides word processors, there are a number of other tools that are useful for writers.

- **Page-Layout Programs**

Most companies have graphic artists to put your work into a final form, but sometimes you may have to do it yourself. If you're contracting to provide a camera-ready product you may want to save the money a graphic artist would cost you (often a mistake) and do it yourself. In any event, if you're working with graphic artists, the more you know about their tools, the better you'll be able to communicate and work together.

- **Flowcharting Programs**

A nice, simple flowchart can help explain complex processes. Some word processors have built-in or add-on flowcharting, and there are a number of fairly low-cost, useful stand-alone flowchart programs.

- **Email**

Whether you're communicating within a company or between locations, email is asynchronous communication at its best. Know how to use it and how to attach and retrieve documents.

- **Presentation Software**

Sometimes teams, groups or companies need to put on a dog and pony show. As a writer, you may be asked to put together a presentation. Presentation software, whatever its shortcomings, speaks the language of board meetings.

- **Spreadsheets**

For huge tables, tables with calculations, and for charting and tracking projects and progress, spreadsheets can't be beat. They're also handy for budgeting projects.

- **Scheduling Software**

If you have a lot of meetings and deadlines, a scheduling program that reminds you when it's time to go somewhere or do something can save your life. Well, at least your job.

- **Project-Management Software**

If the project is big enough and you're managing enough people, you may need project-management software. Of course, once you reach this level of management complexity, you're likely a full-time manager and not writing anymore. Project-management software can be a black hole for time, requiring constant care and continual updates. Before you invest your time and money on a high-end package, first try using a spreadsheet, a wall calendar, and a simple Gantt-chart drawing program.

- **Graphics Programs**

Most writers aren't expected to be visual artists, but you may have to crop, edit and touch up graphics now and then. If you're working in software, you'll definitely have to be able to take screenshots, crop, edit and otherwise prepare them for layout.

- **Help System Creators/Compilers**

If you're creating help systems for Windows-based software, you'll need to know your way around one of the systems for creating and compiling the help file. There are many shareware and commercial products that make this job fairly easy—unless you want to add all the bells and whistles.

- **Web Publishing Programs**

If you write for the web, or you want to supply a help system that can be read with a web browser, you'll need to know how to create web pages and links.

You can use anything from text-based HTML editors to visually based programs similar to page-layout tools to huge, fancy-schmancy web-design programs that include many project management and database features.

HTML is becoming very popular for help systems these days (because of its cross-platform compatibility), as well as for publishing all manner of information on the web.

- **Prototyping and Multimedia Programs**

HyperCard, Director and other software tools are very useful for prototyping interfaces of both software and hardware products.

- **Reference Books**

Old technology, but they still work.

- **Wall Calendars and Whiteboards**

Wall calendars for scheduling and whiteboards for thinking.

- **The World Wide Web**

These days the web is actually a useful research tool. Besides general information on almost any subject, there are a number of sites that have specialized dictionaries and other useful information for writers.

- **Pencil and Paper**

Agonizingly slow, and in my case nearly illegible, but there are times when nothing beats it. At meetings it's quieter and more dependable than a laptop for taking notes. Some writers still like to write their first draft by hand. If it works for you, do it. I mostly use paper for quick notes or rough drawings to help me think.

- **Tape Recorder**

Sometimes you need to deal with information faster than you can write or type. That's when a tape recorder is essential. It's also useful when you are dealing with people who are busy, or who lose their train of thought waiting for you to catch up in your note taking.

If you are taping, you can concentrate on the meaning of the words, and not on the writing or typing. You can only catch errors and missing facts if you are listening closely, and ask for clarification then, instead of days later after you've gone over your notes three times and the person you need to talk to is out of town.

Of course, the problem with tape recorders is that you have to transcribe the tape, which can be laborious and lengthy. A transcription machine helps by letting you keep your hands on the keyboard and move the tape forward and backward with your foot. It also lets you speed up or slow down the tape as needed. Having someone else transcribe the tape for you is also an option. You may be able to draft an administrative assistant or secretary. Personally, I'd rather do it myself. It takes a long time—at least twice the length of the tape—because I do actual writing along with the transcribing. As I write down a point from the recording, I expand on it right then, while the thoughts are fresh in my mind.[17]

Know Your Tools—But Don't Let Them Control You

Our tools influence the way we work. You can't get around that, other than to get another tool. But don't fall into the trap of letting your tools dictate how you organize your final versions. Just because your word processor makes it easy to make bulleted lists, don't make the whole book in bulleted lists.

It's a Balancing Act

Many of the guidelines in this chapter tell you ways to do things, then warn you not to overdo those ways. Techniques that, used sparingly, can make readers feel comfortable, can make you seem condescending if overused. Other techniques that work for some age groups and levels of technical expertise don't work for other reader groups. You need to understand each audience, and develop a feel for what makes them feel comfortable and what annoys them. It's a balancing act of subtlety and tact.

UnTechnical writing is writing. It's not a formula or template. You can't just plug in facts and have a document pop out. It requires research and practice. Gather a group of readers you trust to look over your work on a regular basis. Listen to what they say and incorporate their suggestions into your writing. But don't overdo it.[18] You're the writer. Use your own judgement about their judgement. Once you start to get a feel for the audience, trust your hunches. Then test your decisions on a sample audience.

---

[17] *You can now buy a combination package that contains both a digital voice recorder and speech recognition software. I haven't tried it yet, but probably will.*

[18] *I told you so.*

Tricks and Techniques

Keep It Simple

Simple in language. Simple in style. Simple in organization. Keep Occam's Razor[19] handy, and use it often.

This isn't Scrabble—you don't get more points for long words. You aren't paid by the syllable. Your job isn't to test or challenge your reader, or impress the reader with your vast vocabulary. It's to get the important information from the page or screen into the reader's brain as quickly, easily and enjoyable as possible.

There is a difference between simplifying your language and dumbing it down. You don't have to reduce your writing to grunts—just to the point of easy, comfortable communication.

> "The first and most important thing of all, ... is to strip language clean, to lay it bare down to the bone."
>
> — **Ernest Hemingway**

> "Poor Faulkner. Does he really think big emotions come from big words? He thinks I don't know the ten-dollar words. I know them all right. But there are older and simpler and better words, and those are the ones I use."
>
> — **Ernest Hemingway**

Be Accurate and Complete

You don't want to give wrong information to someone who's already struggling to understand. And you don't want to leave out important facts—or unimportant facts, if they're useful.

OK, OK, this is obvious. But I had to say it. Or the book wouldn't have been accurate. Or complete.

---

[19] *Occam's Razor is the rule that if there are two ways to do something, the simpler one is the better one.*

Be Consistent

There are a lot of ways to be consistent—and inconsistent.

- **Names of Things**

Don't call it a doohickey in one chapter and a whatsit in another. And make sure the name you use in the documentation matches the name on the actual product, and the box.

- **Abbreviations**

When you use them, use them the same way throughout the whole document or series of documents.

- **Level of Detail**

If you're covering a number of subjects, they should all be covered in the same amount of detail—unless there is a reason that the reader needs to know more about a particular subject, in which case that should be made clear.

- **Look and Feel**

This includes the page-to-page and chapter-to-chapter look of a single document, including typography, tables and illustrations, and the book-to-book look of a series of related documents, including typography, overall design, and table of contents and index organization.

- **Presentation of Information**

Like information should be presented in a like manner. Always show the reader instructions in tutorials the same way. Always explain similar things in the same way—if you show a flowchart in one process, then show it for all processes.

- **Lists**

Use parallel style and construction. Note that the headings on this list of ways to be consistent consists of all nouns or noun phrases. If this item's heading was "Be consistent with lists," it would be inconsistent with the rest of the list.

- **Overall Quality**

Don't make half your book wonderful and slap the rest together. And if you're doing a quick cheapie, don't make part of it wonderful—it'll make the rest seem worse than it is.

Working with a good editor is a good way to learn consistency.

Don't Leave Out Obvious Steps

Lots of things seem obvious to us.

Like plugging in the appliance before trying to make it work.

Like taking out the first disk before inserting the second.

Like clicking CANCEL when you want to cancel.

Like clicking DONE when you're done.

As we become more technologically knowledgeable and more experienced with technical products and processes, it is easy to consider basic facts and operations to be obvious to anyone. But these things are not necessarily obvious to the person looking at an appliance for the first time, or who is using a computer for the first time.

What this boils down to is remembering that you, no matter how much empathy you have for your readers, are not your reader. You know too much. It takes diligence—and testing—to make sure the obvious has been covered.

One More Time

Rewrite.

There are very few (if any) writers or writings that aren't made better by a rewrite. Or two or three.

It's nice to think that once something is down on paper (or on disk) that it is done and ready. Or even ready for editing. But it's not.

Trim, simplify, reorganize, rethink, restate. Rewriting is for some the worst, most boring part of writing, but once it's over, you'll be glad you did it. And so will your reader.

> **"Easy writing makes hard reading."**
>
> — Ernest Hemingway

Order of Writing

Unless it's required by your client or an amazingly tight schedule, you needn't write documents in order from front to back.

Usually, the information doesn't come to you all at once or in order. I usually find myself writing the reference section first, starting with whatever parts of the program or product have been locked down and finished, and filling in the rest as soon as the information is ready.

The Introduction section should be written near the end of the project. Since it describes the rest of the document, it makes sense to write it after the rest of the document has at least been written as a complete first draft.

Starting the tutorial(s) before the product is close to being finished is a prime source of suffering for writers. You can outline the tutorial, and identify what you want the reader to learn in it, but don't start writing the steps if the product is still changing.

If you hit slow times, and find yourself waiting for information, then that's a good time to work on background, history or added-value sections of the document.

Order of writing aside, I prefer to turn the whole thing in at once. (Well, at least turn in the final version all at once. Each section will be sent off individually for various preliminary fact-checks and edits.) That way, as I work on one section and come up with something that should have been included in a previous section, I can put it where it belongs, not stick it wherever it fits or slap on an extra appendix. This also lets me jump around while I write; when one section just isn't happening (or holding my attention), I can work on another section. It keeps me writing without forcing.

Of course, sometimes you gotta force.

Sidebars and Other Separated Text

Sidebars and other visually separated chunks of text are a good way to provide related information and deeper detail or definitions without interrupting the flow of the main text or referring the reader to another section of the book.

Typical related information could be a quote that relates to the subject, a joke about the subject, background or historical information that will give the reader a better perspective on the subject, a bit about a famous expert on the subject, or anything that will enhance the reader's enjoyment or understanding of the subject.

For not-so-related, yet important-enough-to-include information, use footnotes.

When writing a sidebar, clearly mark its beginning and end, as well as the heading, for the graphic artist, as in the following example.

[Sidebar: What's a Sidebar?]

A sidebar is a chunk of text, with or without graphics, that is separated from the main body of text, like this. It is typically in a narrower column than the main text and in (or partially in) the margin, hence the name "sidebar." Sidebars can also be the full page width, or even take up a whole page.

Sidebars are often separated from the main text by rules or boxes or other graphic elements, depending on the graphic style of the document.

[End sidebar]

What's a Sidebar?

A sidebar is a chunk of text, with or without graphics, that is separated from the main body of text, like this. It is typically in a narrower column than the main text and in (or partially in) the margin, hence the name "sidebar." Sidebars can also be the full-page width, or even take up a whole page.

Sidebars are often separated from the main text by rules or boxes or other graphic elements, depending on the graphic style of the document.

Writer's Block

One day a friend at work came up to me and said that the woman who works in the office down the hall hates me. I asked why. She replied, "I told her that you're the only person I've ever met that never gets writer's block, and she said she hates you."

I hadn't thought about writer's block in a long time, and realized that it was true … I didn't get it anymore. I sometimes can't write if there is too much noise or too many interruptions. I can be too burnt out or thinking too much about something else. I can be lazy and lack the willpower to force myself to work. And sometimes I procrastinate because I don't like the subject I'm supposed to be writing about. But I can't remember the last time I had writer's block.

It all began many, many years ago with a screenwriting class I took through the UCLA Extension. The class was taught by Milt Rosen, a veteran screenwriter with numerous sitcoms and variety specials (and joke books) to his credit.

Milt was writing something on the board, I don't remember what. Someone in the class took the opportunity to ask a question. "What do you do when you get writer's block?" Milt calmly turned to the class, and said in a very matter-of-fact voice, "There's no such thing," as if that explained everything, and returned to his board-writing.

You have to understand that this was a class of motivated screenwriter wannabes, all of whom had spent untold hours staring at a blank page or computer screen. They had come to this class with high hopes of becoming professional writers. And the local guru just implied that a "real" writer wouldn't ever have writer's block. The class began to rebel: first moans and murmurs, then threats. Milt, with the excellent timing of a master comedy writer, waited until the guns were loaded and the noose was knotted before he turned back to the class.

First, he repeated himself, "There is no such thing as writer's block." Then he went on, "There is only failure to make a decision." At this point, the class was silent, hanging on every word. He went on to explain that whenever we're stopped in our writing, it's because we fail to decide what happens next. Whether it's what Johnny does now that he's hanging by one finger from the 132nd floor of the building with the mad slasher fondling his axe and eyeing Johnny's last gripping finger—or how to describe what happens when you press a button, all you have to do to move forward is make a decision.

So when you're stuck, make a decision. Any decision. Then write. If it doesn't work, go back and change your decision. With today's word processors, it's easy to change things. And you're not wasting any more time changing something than if you sat there doing nothing, waiting for that perfect line, thought or phrase to come.

It works for me. Try it. Maybe that woman in the office down the hall will hate you, too.

Jump Starts

Even though I've said, and firmly believe, that there is no such thing as writer's block, there are times when you can get bogged down. It can be a sentence that just won't cooperate, or a paragraph that just winds down to nothing. At these times, I have a couple of tools to help push through the bog.

Changing Perspective

Have you ever written a sentence that didn't work, and tried and tried to tame it, but it just got more and more convoluted and confusing? When this happens, step back. Stop trying to write the sentence, and ask yourself this question, "What am I trying to say?" More often than not, your answer to this question will work better than all your previous efforts. Changing perspective is a wonderful tool.

The Running Start

Sometimes it's hard to get started, even in the middle of a project. That's when I get a running start: I back up a few pages and read it through, sometimes editing along the way, sometimes not. By the time I get to the new day's starting place, I just keep going.

Humanize Technology—and Yourself

The traditions of technical writing and journalism frown on mentioning yourself or your personal experiences. You'll see an occasional mention of "this author" or "the author" but rarely an "I" or a "me."

Well, get over it. There's no need for that fly-on-the wall stuff in UnTechnical writing. This isn't journalism.[20] We are tour guides, taking our readers with us on a journey. We're there with them, holding their hands and leading them to the promised land of understanding. We're technological Virgils, guiding the newcomer through a treacherous domain.

Don't be a nameless, faceless cog in the machine that cranked out the product. Let the reader know that you're real and that you care about them and that you tried your best to make their experience a good one. It's OK to be a person, to let the readers know that an actual human, who went through the same learning process as they are about to go through, is writing this to make their lives easier.

---

[20] *For great examples of journalistic fly-on-the-wall nonfiction writing, read anything by Tracy Kidder, especially* Soul of a New Machine. *Great writer, great books. But ... maybe it's because of my technical background, when I read his books I find myself wondering where he was, what he was doing, and how much effect he had on the situation just by being there. In any event, you've got to admire Mr. Kidder for creating the perfect job for himself: watch other people work for a year or so, write a book about it, win a Pulitzer. Where are those job applications?*

Let the reader share your discovery process. Develop the attitude, "It's you and me against this thing, and since I've been through the wars already, I'll help you out." As an example, look at Carlos Castaneda's writings. He used himself as the foil, the thick-headed goofball, to offset the wisdom of Don Juan. The guy wasn't stupid. He used this as a way to help the reader feel comfortable with new ideas, to help them realize that "Hey, the guy who wrote this book was as shocked and confused as I am right now, but he figured it out eventually, and so can I."

The trick here is know the limits. Don't turn the document into a biography or self-exploration novel. The product is the star. Don't steal the scene. That fine line between being a guide, a presence, a friend and a scene-stealer takes time, practice and rewrites. Find a good editor and a group of readers that will give you honest feedback.

Of course, like all techniques, don't use it all the time—only when it will help the reader. Don't overuse it. And, above all, don't do it better than I do. We may be competing for the same gig someday.

Show Your Love of Science and Technology

If you really care about the subject, let that caring through. It's OK to get excited about something wonderful. And it's more than OK to get the reader excited about the subject.

I gained my appreciation of and fascination with science and technology through reading science fiction. Sure there was story there, and the science wasn't (yet) real. But I felt the fascination the writers had with technology. It was something to look forward to in my own life. Why shouldn't the UnTechnical writing of today pass on some of that same thrill—"Here is great new technology, and it's a real part of our lives. Learn it and use it. Enjoy it."

This is common in today's popular science books, and can be applied to product manuals and other technical documents as well. But don't overdo it on the product side. You're not selling the thing—they've already bought it.[21] (But don't hold back too much—you *are* selling the next thing.)

---

[21] *An example of overselling is the packaging for Zip disks. Zip drives and disks are wonderful products. But their marketing department is a little overzealous. Of course, when you go to the store and see a pack of disks on the shelf, there will be product promotion on the outside of the package. Fine. And when you get it home and take the disks out of the package there's*

"Nothing astonishes men so much as common sense and plain dealing."

— Emerson

Write the Way People Talk ... Sort of

Have you ever been totally stumped by a technical document, then asked a friend for help? And he or she explained it in nontechnical and possibly ungrammatical terms, but it got the message across? Why did that work? Partly because it came from a friend and partly because it was in your spoken language—two factors that make you more comfortable.

Your goal is to get the information from your mind into the reader's mind as quickly and painlessly as possible. So write the way people talk. Go ahead. End sentences with prepositions. Use slang. Use conversational grammar. When readers feel comfortable, they absorb more.

Avoid sounding academic at all costs. You're writing for your friends, your neighbors, even your mom—not your professor. There are many people who will turn off to the subject and your writing if they feel like they're back in school.

But be careful—it's easy to overdo this. If you write exactly the way people talk, there'll be a lot of ummms, aaahs, pauses, abrupt changes of direction, and a lot of confusing blather. And slang gets old fast. (Twenty-three skidoo!)

Write the way people talk—after they've been edited, rewritten and well-rehearsed by a great dialog coach.

"Write with the learned, pronounce with the vulgar."

— Benjamin Franklin

---

more product blurb on the insert card in each individual disk case. OK. But when you open the individual disk case and take out the disk, there's more promotional blurb on the inside of the insert card. OK already! I get it! Get off my case! I know what it is, I already bought it. Stop trying to sell it to me! Blank media, from audio cassettes to data backup tapes, always came with that little insert card. And at least one side of that card was blank, so you could write on it to list the contents of the media. Convenient and useful. But by plastering both sides of the card with marketing slogans, this potentially useful card becomes nothing but landfill.

Sexist Language

Think about your audience. If it's mostly men, then you can use the pronoun "he" freely and exclusively. If it's mostly women, then you can use only "she." If, as usual, it's a mixed audience, then you have the potential to annoy your readers. While we UnTechnical writers may want to surprise or even shock a reader now and then, we don't want to annoy.

When you find yourself in the pronoun dilemma, you have a number of choices:

- Use only "he."
  Using "he" as a pronoun to refer to an individual who may be either male or female (but you don't know for sure) is the traditional English usage, dating from before political correctness. It isn't technically wrong, but it can make some readers think you're a jerk and pay less attention to you.

- **Use only "she."**
  On the opposite end of the scale, you can take the attitude that women have had to put up with being called "he" for a long time, so let's turn the tables and let the men see what it's like. While fair, and technically legal, it can still alienate a good portion of your audience, both male and female.

- **Alternate between "he" and she."**
  This is the "fair" approach. Approximately every other time you use a pronoun to refer to a person whose sex you don't know, switch between "he" and "she." While it's fair, legal, and generally acceptable, it is also noticeable. The reader is very aware that the writer is switching back and forth. This can be distracting (at least for me); the reader stops thinking about the subject and thinks about the switch.

- **Use "he or she."**
  If the writer wants to cover all the bases without switching, he or she can use "he or she" each and every time. Again, legal. But there's something ... wishy-washy about it. And it's longer and a bit more awkward. I may use this once in a while, but not often, and rarely more than once in a chapter.

- **Pluralize and use "they."**
  My usual way around the whole issue is to take advantage of the fact that in English, the plural pronoun, "they," works for both sexes. Instead of writing something like:

  If the reader likes the subject, then he or she should read the book.

  I'd write:

  If readers like the subject, then they should read the book.

  At least 80% of the time you can pluralize and use "they" to avoid sexist—and awkward—language without ruining your point. In those few cases where pluralizing just doesn't work, then you can fall back on one of the other options, or … just use "they."

- **Just use "they."**
  Many editors and grammar mavens are now allowing the use of "they" as an any-gender singular pronoun for technical, if not for academic, writing. A reader may find that they are referred to as "they" even though they are one person. Most readers won't notice, but some editors are still sticklers. I use it over "he and she," but not very often.

- **Get cute.**
  A last possibility is to use some sort of combination of he and she, such as he/she or s/he.[22] These work, but they seem like the writer is trying too hard. Maybe, if and when they come into more common usage, they'll lose their cuteness.

Write in Bite-Sized Chunks

Keep sentences, sections and paragraphs small—write in bite-sized chunks. This forces you to be clear, to keep your explanations simple and stops you from rambling on too much.

And it keeps the reader reading longer.

---

[22] *I've heard it said that even the combination word "s/he," which means "she or he," is showing undue prejudice against nonsexual entities, and the proper combination word should be "s/he/it." Of course, it would never work in polite conversation.*

Keep the Reader Reading

Anything we can do to keep the reader reading is good. Anything we can do to help them get through it quickly and painlessly, we should.

Most people don't read technical books and manuals for pleasure. They only do it because they have to. And they're looking for any excuse to stop, to take a break, to get a cup of coffee or to throw the book across the room in frustration. A long, densely written paragraph is just such an excuse.

A small, simple-looking paragraph smiles at the reader and says, "I'm quick and easy. Come on and read me, you'll be done before you know it." A large paragraph says, "I dare you to read me, and to try to wrest meaning from the dark, twisty caverns of my many long sentences."

The end of a chapter or section is another good excuse to stop. So make the first paragraph of each chapter and section a small one. If it's inviting enough, the reader will put off stopping until the end of the next chapter or section.

The quality of the layout can also help keep the reader reading. Lots of white space on the page and a good space between paragraphs makes the page look inviting. A page crammed full of solid text looks scary.

Use Lots of Headings and Subheadings

Using a lot of headings and subheadings serves four purposes:

1. It keeps the book divided into small, easy-to-absorb chunks.

2. It allows the reader to zero in on the exact section they want to read without fishing around.

3. It familiarizes readers with all the sections—as they look for the topic they want, they read all the other headings. When they need to look up something else, they'll often remember that there was a heading about that subject.

4. It makes it easier for you to organize the book and explain each subject independently.

A Little Humor Goes a Long Way

By this I mean both that some humor is good and that too much is bad.

People like humor. There are jokes and comics in just about every magazine and newspaper. Humor relaxes us. We learn better when we're relaxed. Think back on your school days. Did you ever have teachers who took their subject so seriously that it bored the hell out of you, and even worse, made you scared to make a mistake or "play around" with the subject or related ideas because you might be treading on a sacred cow? Did you like that subject? Did you learn very much?

Compare that with the teachers (if you've been lucky enough to have any) that included humor in their presentations. These teachers didn't take the subject so seriously that it hurt. You were allowed to play around with the ideas.

Humor defuses a tense situation. If you can laugh at it, you're no longer afraid of it. Many readers are afraid of the technology you're writing about and afraid of breaking it and afraid of how it will change their lives and afraid of having to read your manual. They need a good laugh.

> "Humor's the true democracy."
>
> — **Robert U. Johnson**

That said, here are the inevitable caveats....

Humor Caveat #1: Know when not to use it.

While I personally can't think of a subject that couldn't use a little lightening up, maybe you can. Don't force jokes about something that you personally can't comfortably joke about. Even more important is how the reader, and your client or employer feel about the jokes. Some people, including bosses, have no humor. Know when to keep your humor to yourself.

> "I never dare to write
> As funny as I can."
>
> — **Oliver Wendell Holmes**

Humor Caveat #2: Test your jokes.

Think about the funny people that you know. Are they always funny? Are they funnier in public when they're really trying to impress someone than they are when they're just hanging around? Probably. That's because around close friends or small groups they're testing their material. Maybe intentionally, maybe subconsciously, but they'll try things out with friends to see the reaction. If it's good, then they'll add it to their repertoire and use it at a party. If not, they'll either drop the joke, or rework it and try it again. It's market testing with friends and family as focus groups. With very few exceptions, the famous comedians you see on TV aren't as funny around their friends and family because they're trying out a lot of untested stuff.

Humor Caveat #3: Ask for help.

If you don't have—I won't say *good,* but how about—a widely accepted sense of humor, get someone else to supply a few jokes or comics. There's bound to be someone around the office that can help. If you have a big budget, there are services that will sell you humor. And if you look hard enough, you can find local amateur comedians and cartoonists that would love to have their stuff published for free or cheaply just for the exposure.

Humor is a very subjective thing. And it's not easy. It takes years of trial and error to find those thin lines between funny and stupid, hilarious and disgusting, uproarious and insulting.

> **"Men will confess to treason, murder, arson, false teeth, or a wig. How many of them will own up to a lack of humour?"**
>
> **— Frank Moore Colby**

Humor Caveat #4: Watch your mouth.

Unless the product you're working on is of an "adult" nature[23] or intended for adults with an adolescent sense of humor, stay away from insults and bathroom humor. Again, it's important to know your audience to know your limits, but when in doubt, go the tasteful route.

---

[23] *Remember when "adult" meant* grown-up? *These days it means X-rated.*

"Guess his humor ain't refined
Quite enough to suit my mind."

— Ellis Parker Butler

Humor Caveat #5: Humor is often a cultural thing.

Much of humor is culturally based, so what's funny to you may not be funny to someone in Japan or Germany or France. Some subjects that we casually joke about here may be seriously taboo elsewhere. If you know that what you're writing will be translated to other languages, be prepared for your humor to be chopped out. Make sure this chopping won't affect the usefulness of the document.

Humor Tips

Jokes and puns can be very effective in headings. If they're good enough, the reader will turn each page of the book and read all of them. But don't let the joke prevent the heading from serving its central purpose of explaining what the section is about.

When the reader looks through the table of contents and sees all the headings, they have to be able to make sense of them and know what they're about. When they want to know what the History Eraser Button does, they need to be able to find something very close to History Eraser Button in the contents, and not have to decide between looking at the pages for "Where Has the Time Gone?" and "Like, I'm History, Man." Make it easy on the reader. You have a lot more leeway for humorous headings for different sections of tutorials than in reference sections.

For chapter or major section headings, you can use two-parters: a joke line and the technically accurate line separated by an em-dash or colon.

Occasionally, you may be able to insert some humor right into the main body of the text. It can be a reward for those who actually read the doc, or it can be like the parrot scene[24] in *Citizen Kane,* and just exist to wake up the reader. Keep these in-text humorous lines and

---

[24] *Near the middle of Orson Welles' Citizen Kane, considered by many the greatest American film of all time, there was a two-second scene—a close-up of a loudly screeching parrot. Critics argued over the deep meaning and significance of this scene. Orson Welles claimed that its sole purpose was to wake up the audience in a slow-moving section of a long film.*

asides very short and to the point. Lighten the mood, but don't slow down the reader or change the subject. Avoid non sequiturs; they're confusing.

Cartoons can be used to get a point across, as evidenced by the political sections of newspapers and magazines all over the world. You can use cartoons to send an important message to the readers without seeming to preach.

Sprinkling little jokes, puns or tiny cartoons throughout the book may get readers to browse the book through at least once. (The first thing I do when I look at a magazine is flip through every page and read all the cartoons.)

An ongoing story or comic spread through the book is another way to encourage (bribe) readers to look at every page at least once.

Page Flippers

In addition to comics and jokes, you can also use quotes, tidbits of relevant information or anything else that's related to the subject and interesting to get the reader to flip through all the pages.

One possible exception is putting flip-the-page animations in a page corner. They're really cool, and people will flip through them, but the book stays so closed and the pages flip so fast that they don't serve the purpose of getting the reader to look at each page.

Don't Overuse Cross-References

I just don't like a lot of cross references. Once in a while, they're very useful, both for the writer and the reader, but when overused, they're annoying.

Have you ever read a manual that had cross references on every page? Did you look at them all? If you did look at them all, did you lose your train of thought? (This is actually the same beef I have with too many hypertext links.)

I use them in this book, mostly to point to the Recommended Reading or related sections, so you, the wonderful reader, will know that they're there when you need them. Not because I expect you to jump there and read something and come back and continue on as if you've never left.

One reason to use a lot of cross-references is to avoid duplication of information. It has been written that you should never duplicate information in a technical document. The reason for this is to make your job easier: if information changes, you'll only have to update it in one place.

This is one of those areas where I diverge from the generally held beliefs of technical writers (and writers of books about technical writing). I avoid cross-references whenever I can and duplicate information whenever it makes the reader's job easier. So I spend an extra hour or two per revision to make sure I don't miss anything. No big deal. If you're really worried about it, keep project notes that list the duplications.

Don't be afraid to repeat. I repeat, don't be afraid to repeat. We—and our readers—learn through repetition. So don't be afraid to repeat.

Hypertext

This book won't go deeply into hypertext, which includes HTML, Windows Help, and other, similar things. But putting too many links or jumps on a page—unless it's a contents or index page—is annoying. If you're compulsive and want to read and know everything, then you jump so much you lose continuity. If you get annoyed by all the jumps, you skip them and miss a lot. If you calmly pick a few jumps here and there, you still miss a lot.

Use hypertext jumps like section references: to lead to related or more detailed sections, not to hold information that should be included in the page with the jump.

And don't believe that just because you are using hypertext that you don't have to organize your writing into a linear or near-linear piece. You still have to do it, and provide the reader a way to move linearly through the majority of the document.

Onscreen Text

Writing text for onscreen reading is different from writing for paper. For many, it's harder to read from a screen and most people, even those who work at software companies, just don't like reading a lot of text on a screen. It ties the reader to their computer and the reader's hand to the mouse or keyboard. Keep onscreen text short, to the point, and heavy on the procedures, as opposed to background explanations.

Printed docs can be set next to the screen and readers can quickly move their eyes between book and screen. But when the instructions are on the screen, they will often obscure the part of the program they are describing. So when you force the customer to leave the program for onscreen help, keep it short and to the point, so they can get back to what they were doing quickly.

The worst onscreen text is a non-interactive tutorial. Multiple pages of steps and results and explanations for a program that can't be seen because the tutorial is blocking their view. On the other hand, a well-designed, well-integrated interactive tutorial that brings up a small window that doesn't block much of the screen can be wonderful for the customer. Of course, it takes more time and money, and requires a lot of programming work.

Laying out text and graphics for reading on the screen is different from laying out for the printed page. The screen is usually smaller than a full sheet of paper, and the text must be larger.

For financial and practical reasons, more and more docs are being put on disk and onscreen. In general, customers don't like it,[25] but time and money will win out.

Now that 17-inch and larger screens are starting to become common, long onscreen docs will be a little easier to use, but won't be fully accepted or totally useful until everyone has bigger or multiple screens at a higher, easier-on-the-eye resolution.

---

[25] *While at Maxis, I asked the customer support group to do an informal phone survey on onscreen docs with customers who called in with questions. As I say, it was informal. It only included about 100 customers, and they weren't a scientifically selected group. But the vast majority didn't like onscreen docs. They found them intimidating because they didn't know how to use them. They preferred to have a printed manual that they could hold and page through at their leisure. And more of the people in the survey said they would call technical support or ask a friend for help before using onscreen help.*

Process

Use Lists

As technology writers, we deal with a lot of information, a lot of details and a lot of steps. It's easy to forget things, lose track of details or skip steps. One way to keep things from falling through the cracks is to use checklists for various procedures.

I devised a number of procedural checklists with the writers that worked with and for me. They include:

- Writer's Overall Project Checklist
- Manual Writing Checklist
- Quick-Start Guide Checklist, and
- Editing Checklist

We used these lists as tools, not as rules. We usually modified them for each project after spending a little time with the producer or project manager and getting the specifics of the project.

These checklists can be found in the Exhibits section at the back of this book. Most likely, you'll need to modify them on a project-by-project basis before you can make full use of them.

Besides procedural checklists, there are other types of lists that come in handy. Unless you know everyone on the project well, make up a contact list of everyone on the project, their title/position/project responsibility, and their phone, fax and email numbers or addresses. If you're working in a different location, list their mailing address as well.

If you work at a company or regularly contract there, you might also want to work up a generic deliverables list (all the things that you as a writer normally deliver on a project). A modification of this list will help you estimate your time and charges. A sample deliverables list is supplied later in this book. It will let you track actuals against estimates, giving you more accurate information for estimating the next project.

Define the Project

Product development can be a huge, lengthy, detailed process, and often details, such as what the writer is expected to do, are overlooked, forgotten or remembered in creative ways.

Before you begin writing on a project, make sure your responsibilities are clearly defined, and in writing. Hundreds of questions will pop up at some time during the project. What are you writing? Due when? Are you responsible for layout as well as writing? In what format should the document be delivered? Are you responsible for graphics? If it's software, what about a help file? What format? Compiled or raw data? The more answers you have up front, the better.

If you're a contractor, then chances are you'll have something in writing that explains what you're expected to accomplish and by when. Depending on who you're working for, you may have everything, down to the word count, in the contract. Or it may be more informal, and the exact number of pages or even number of books, help screens, and quick-start guides that you are expected to produce may be up in the air. Adding work to your contract is fine, as long as time and pay increase as well. Read carefully before you sign.

Even in the situation where you're an employee of a company, you'll need a definition of your proposed tasks that you can base your time and cost estimates on. And having it in writing is best for all concerned.

For this purpose, my group devised an in-house Writing Request Form, which is included as an exhibit later in this book. Over a few years and many projects, this form was rarely completely filled out and signed by all, but it still served the purpose of making the producer or project manager aware of the scope of the writing on the project, and let them know that they had to think about this stuff and eventually make some decisions. It also served to cover the writers when they were asked why they were late delivering something they had never heard of before.

Glossaries and Inserting Definitions Into Text

Glossaries are good, but don't make the reader stop their reading to look up a word more than once every couple of pages.

If you have to use a lot of words that are technical or new to the customer, then try to insert a definition into the text, either as part of the main sentence or in parentheses.

Here are three examples, showing how you can define a term within the text:

Ants use trophallaxis, or the sharing of vomit, to spread food and information.

The process of spreading food and information through the sharing of vomit is called trophallaxis.

Ants use trophallaxis (the sharing of vomit) to spread food and information.

If this ruins the read, makes the text too long or awkward, or just doesn't feel right, then put definitions into sidebars or at the bottom of the page in footnotes so the reader can look them up without having to flip around the book. *Cricket,*[26] a children's magazine, is a good example of putting definitions in the margins, along with pronunciation help.

If there are more than a half-dozen or so words you need to define for the reader, and they appear in more than one place, then include a glossary—even if you already defined the words within the main part of the document. If the reader forgets a definition by the time they see the word a second time, they can use the glossary, and not have to search for the last place they saw the word.

Notes Within the Text

There is a saying in publishing that you shouldn't ever put anything into a manuscript that doesn't belong there and that you wouldn't want to see printed. It's probably a good, safe idea, but the convenience of being able to leave notes to yourself, questions for technical reviewers and editors, and placeholders for graphics or text additions is too much to resist.

---

[26] *Cricket is an example of a children's print product that is user friendly and is designed to make it easy for the reader to read, understand and learn. We never see that in magazines for adults. Too bad. The learning process never stops, and anything within reason that makes it easier is worth doing.*

As a safety precaution, make sure this "extra" text is well-marked for easy removal. Above all, don't use flippant, inappropriate or off-color language that you expect to clean up later, especially loosely sprinkled throughout the text.[27]

If you put notes or comments in that you wouldn't want to see printed in the final version, make it your responsibility to see that it's all cleaned up before it's too late.

I use brackets ([ ]) in my drafts to indicate non-reader material. I chose brackets for this task because I rarely use them for other purposes in documents. If you use brackets often, then choose something else to mark your questions and comments. Just make sure that what you use won't confuse yourself or others, and that your word processor can do a search for it.

I also bold the brackets and the text inside them to make the note or question stand out. This makes sure the question will be answered, makes the bracketed text easy to find when it's time to remove it, and distinguishes it from those few times when brackets may be used within the document.

Some typical examples of in-text notes that I commonly use:

- **Callouts for graphics**

[g21—Screenshot of Save Dialog. Caption: The Save Dialog Box]

Putting the caption into the callout helps make sure the right caption goes with the right graphic, and makes it handy for the graphic artist to copy and paste the caption where they want it without retyping.

---

[27] *I almost learned this lesson the hard way. I hadn't even finished a very rough first draft of a science section to accompany an educational computer game about ants. In the pre-draft, I had referred to insect reproduction a few times by other, less polite terms. My excuse … uh, let's see … uh … after a couple of months of studying and writing about ants, you can get a little light-headed, or the pheromones made me do it. In any event, during a visit from our Japanese publisher, someone grabbed the draft from my computer to give them for a preliminary translation. Luckily, I found out in time and did a quick search and replace before the document left the country. On the other hand, I might have had some interesting conversations with the translators—or, if they didn't catch it, I might have received some irate calls from Japanese entomologists.*

- **Notes to myself**

[more here] or

[recheck all button names in this chapter after the final art is in the product] or

[This section will describe every Edit Window button and function. Should be ready July 22.]

Especially in software, where you often begin the manual before the product is completed, you work on the parts that are done and finish each additional section as the program is completed. Leaving notes to myself within the text about the unfinished parts helps keep me from missing something. Since the notes are easy to find, I'll often copy and paste them into another document, creating a list of unfinished sections for planning and scheduling purposes.

- **Comments to graphic artists**

[David: please make this graphic as big as possible on the page, it has a lot of detail that needs to be seen]

- **Questions for editors or technical fact checkers**

[Programmers: Did I explain this process in enough detail to be accurate? Tech support: Is this too technical and/or boring for our customer base?]

Put people's names and/or job titles right into the question if you know who should answer it. It catches their attention.

Be sure to remove all brackets and comments (except those needed by the graphic artist during layout) as part of the final pre-layout edit. Do a search for left and right brackets to make sure you get them all before passing on the draft. Also, encourage the graphic artist to do another search for brackets near the end of the layout process.

Graphics and the Graphics List

As you write, self-edit and rewrite your document, you insert place-holders, or calls, for graphics in the places where you want graphics to go. Typical calls for graphics would be something like:

[g103 of Cancel button]

or

[g007 of main screen, blank document showing. Caption: The Main Screen. Callouts: Menu Bar, Toolbox, New Document][28]

Notice that the graphics call contains:

1. a "g" to let you know that this is a call for a graphic rather than a note,

2. the name of the graphic (the number), and

3. the captions and callouts, if any, for the graphic.

The graphics call is also contained in brackets, which are bolded, both to make them stand out so they won't be missed during layout and to visually distinguish these brackets from any other brackets that you might use in the document.

I still use numbers for graphics file names, even in this day and age where you can give files nice, long descriptive names. Having the graphics numbered keeps them in order and makes it easier for you locate and check them, both in your document and as separate files, and much easier for the graphic artist to link or insert them into the final document.

While writing, I don't name (number) the graphics. I just put in a call that indicates the graphic and its contents, like:

[g of Cancel button]

Then, at the final draft, when I'm fairly sure that all the graphics have been specified, I go through the document from beginning to end, and name (number) all the graphics in order. If by chance a graphic needs

---

[28] *Use extra zeros before the graphic numbers so all the numbers contain the same number of digits. Once you know roughly how many graphics you have in the document, you'll know how many zeros to start with. This keeps the numbers in order even though computers alphabetize directories. Otherwise graphic number 2 would be after all the 100's, and 3 after all the 200's. A VCR manual may only have a dozen graphics or so, so this wouldn't be so important, but a piece of software with lots of buttons, dialogs and windows can have many hundreds of graphics.*

to be added later, then it gets the number of the previous graphic, plus an "a," so there might be a g202a.

After the graphics have been numbered, I create the graphics list. I use this list as my guide when creating the graphics and taking and editing screenshots. It's easier than paging through the hundreds of pages of the original document, and much better for tracking progress and locating late or problem graphics.

To create the list, I copy all the graphics calls to another document, turn that list into a table, add a few columns and it's ready to go. It can save a lot of time if you write a macro to search your document to find all the bold left brackets followed by a "g," copy the whole line, paste it to the end of the new document, then have it remove all the left brackets and the "g."

Graphics list step 1: copy all the calls for graphics to a separate document.

[g01 of OK button]

[g02 of Cancel button]

[g03 of History Eraser button]

[g04 of Self-Destruct button]

[g05 of Main screen, blank document showing. Caption: The Main screen. **Callouts:** Menu Bar, Toolbox, New Document]

[g06 of Load dialog box screen, blank document showing. Caption: Load File dialog box. **Callouts:** Select the file from this list, Select the disk from this list]

[g07 of Save dialog box, blank document showing. Caption: Save File dialog box. **Callouts:** Select the target disk from this list, Name the file here]

Graphics list step 2: Get rid of the extra brackets and letters.

01 OK button

02 Cancel button

03 History Eraser

04 Self-Destruct button

05 Main screen, blank document showing. Caption: The Main screen. **Callouts:** Menu Bar, Toolbox, New Document

06 Load dialog box screen, blank document showing. Caption: Load File dialog box. **Callouts:** Select the file from this list, Select the disk from this list

07 Save dialog box, blank document showing. Caption: Save File dialog box. **Callouts:** Select the target disk from this list, Name the file here

Graphics list step 3: Turn it into a table.

Name	Description/Captions/Callouts	Status	Notes	I18N	GA Use
01	OK button	done			
02	Cancel button	done			
03	History Eraser button	done		✓	
04	Self-Destruct button				
05	Main screen, blank document showing **Caption:** The Main screen **Callouts:** Menu Bar, Toolbox, New Document	FPO	See attached for callout placement		
06	Load dialog box screen, blank document showing **Caption:** Load File dialog box. **Callouts:** Select the file from this list, Select the disk from this list		See attached for callout placement		
07	Save dialog box, blank document showing **Caption:** Save File dialog box. **Callouts:** Select the target disk from this list, Name the file here		See attached for callout placement		

The table headings you use in your graphics list will include:

- **Name**—the graphic's name/number from the graphic call, with the "g" removed.

- **Description/Captions/Callouts**—the rest of the information from the graphic call. If you like, you can separate descriptions, captions and callouts into separate columns, but it's generally not worth the effort, and would require very wide paper or very tiny fonts to print it.

- **Status**—this is where you can keep track of whether the graphics are done, partially done, problems or note that there are temporary placeholder graphics, called FPO: *for position only.* Sometimes the text will be done and the layout must begin before the final graphics are ready. If you supply the graphic artist with a placeholder that's the correct size, the layout can proceed, and the final graphic can be dropped in at the last minute. Just be sure to put FPO on this list and, if possible, in big letters right on the placeholder graphic itself.

- **Notes**—This column is for any other information not related to the graphic. Notes can be to yourself, like, "This dialog may be deleted from the program," or to the graphic artist, "Please make this as big as possible on the page."

- **I18N OK**—A checkmark here means that the graphic may be used in versions for any language; generally a graphic with little or no text. (I18N is the abbreviation for internationalization, which means to make foreign language and foreign culture versions of a product. More on this later.)

- **GA Use**—this is a courtesy for the graphic artists. It gives them a place to track the steps, status and problems as they work on the graphics. As long as you're giving them the list to read your notes, you might as well make it useful for them.

Once the list is ready, I print it out, then go through it to identify all the easy graphics to do first. I also look for duplicates: graphics that I can create once and save under two or more names—this happens a lot with buttons in a software program.

As I work, I update the list by hand. Every so often, I update the electronic copy and print out a new one to work from.

Once the graphics are all taken (or at least most of them), I print out each of the graphics that have callouts, write in the callouts and draw lines to the part of the graphic that they describe. These printouts accompany the graphics and list when they are turned over to the graphic artist for layout.

When Size Counts

Most of the time, you'll be asked to make the manual shorter and smaller, for cost reasons: writing costs, editing costs, layout costs, printing costs, paper costs, shipping costs.

One way around this is to print everything very small.

Don't. You can fit a massive tome into a 24-page CD insert by using tiny type, but nobody will be able to read it, which makes it worse than useless.

To make it smaller (if you absolutely have to), cut the least important information. It hurts, but unless you're willing to pay the additional cost per product, sometimes you just have to do it.

Sometimes (miracle of miracles) someone, possibly from marketing, will come to you and say they want the manual to be bigger. This may be to make the product seem more impressive and important, or just to make it heavier, so when the customer picks up the box, it has some heft to it and feels like it's worth the price.[29] (You may be able to use this to your advantage as a reason not to cut the tutorial section in half. "It'll give the product more perceived value!")

---

[29] *This is especially true with software, where the product can take years and millions of dollars to create, and it all fits on a CD that can be mass-produced, case and all, for well under a dollar. The customer has the tendency to pick it up and think, "What? $50 for this? No way." But if there's a big book in the box along with the CD, the product has more heft, and feels more expensive.*

In those instances where you need to add pages, try to add filler that's actually useful, enjoyable or related. Some examples that have worked for me are:

- Historical background information
- Scientific background information
- An art gallery (drawings, essays, poems about the subject)
- Quotes
- Related legends and folktales

> "Not that the story need be long, but it will take a long while
> to make it short."
> — **Henry David Thoreau**

Explaining Through Story

Over the years and the projects, I occasionally thought that it would be great—and great fun—to do a manual in the form of a short story or novelette, with all the information included in the story. Before I actually did it, I ran across a couple of manuals that tried it: a space simulation game and a screenshot utility program.

The space simulation game was a colonize-the-moon game. The story in the manual was about a possible situation that could happen to people while colonizing, and they had to deal with all the same controls that the game player deals with. But the characters weren't in the game, so I didn't care about or identify with the characters at all. I found myself skipping most of the story and skimming for the information I wanted. In some places the information was broken out in screenshots and charts, but some I really had to dig for. My overall experience of this story-based manual: it was better for the writer than for the reader.

The utility program is a Mac-based screenshot taker/editor. (Exposure Pro—a great product for Macs, but there never was a Windows version.) Their story-based manual worked better for two reasons: The story was about me—well, about someone like me who wanted to use this utility to save time and work. It was actually presented in a story form, but it wasted very little time and space on unnecessary details. It introduced the character who wanted to do what anyone using the program wanted to do, described the things he needed to get done for the project, then told, step-by-step, how he used the utility to get the job done. There were only occasional sentences here and there giving others' reactions (all positive, of course) to the work.

It worked. I read through it all. I felt no need to skip pages or paragraphs, since there was never more than a sentence or two at a time that wasn't useful information. What it really was, was a tutorial put into context through the use of a story. Instead of saying, "Click this button and that happens," it said, "Joe clicked this button and that happened." An acceptable variation. Wouldn't want to see it in every tutorial I look at, but a nice break now and then.

The other thing that was good about this manual was that the whole thing wasn't in the story form. Just the tutorial. There was also a storyless reference section including charts of what each menu item, button and function did.

After looking at these two samples and seeing what worked and what didn't, I came up with these guidelines:

- Story is OK for tutorials, but keep the story part short and simple. Go more for a few laughs than for depth of characterization while spooning out the information.

- If you are writing about a game or other product with characters, then use them to give depth to the story, and to get the players into the mood. Don't bring characters into the manual that have nothing to do with the program.

- Keep the reference section (and/or reference card) as clear and clean and unadorned by story as possible.

The most story-like tutorial I've written was for the *SimLife Official Strategy Guide,* published by Prima. I had already written an extensive, very clean and unstoryish tutorial for the product's manual, and wanted to give readers of this aftermarket book something extra, something different. And it was a way for me as a writer to have a little more fun while doing a very complex, difficult job over again.

Going Beyond the Product

Sometimes you have to teach the big picture, going beyond your product to the environment it works in.

The most obvious example is in software, where you may have to give some background in standard operating system use, in addition to your program that works within the operating system.

For DOS, each manual practically needed to be an operating system text. And even with the updated Windows and Mac operating systems, you still may need to explain some basics, including window manipulation, file management, disk use, and even basic mouse and keyboard use.

You can't assume that the customer knows the environment. And that goes for products other than software, too. You may not need to explain the theory of electromagnetism in the instructions for every battery-operated toy, but you may want to say things like:

- Don't get the toy wet.
- If the toy stops working, replace the batteries.
- Only use batteries that meet this description, or use a battery eliminator with these specifications.

Describing only your product, and not providing any of the bigger picture is a sure way to increase technical-support phone calls and product returns.

Let Customer Service and Technical Support Staff Look It Over

Yes, you're the writer, but these are the people who know the customers intimately—and they'll be the ones who'll pay the price for your omissions and mistakes. They should be thrilled to help.

By the time the customer sees the product, the engineers and producers who worked on it are already off to a new project. While you're putting the finishing touches on your manual and asking for feedback and accuracy checks, the people who have been on the project for a year or two are sick of it and are already thinking more about some time off and the next, new project. But customer service and tech support are getting ready for the onslaught of customer calls about the new product, and want and need to learn all they can about it.

If you want real, customer-based feedback and accuracy checks, then you need to stay in touch with customer service and technical support. And if you want to minimize negative customer-based feedback, then get customer service and technical support involved before shipping.

As a minimum, let technical support and customer service check out anything to do with installation or setup—prime sources of customer confusion. Beyond that, leave it up to them—or more accurately, their manager—to decide what they want to see and have time to read and comment on.

In a company setting, always work through the managers. They can assign the right people to give you the feedback you need, saving you hours of explanation and reading of comments that may make no sense. Also, if you mess up schedules or cause a breakdown in the system by handing everyone in the department a 300-page manuscript, you'll make enemies and never get help when you need it.

Usability Testing

Usability testing is often done on a product and its interface, but it's just as important to test assembly instructions, installation instructions and tutorials. And it couldn't hurt your reference section either.

As far as product testing goes, if you as the writer can get involved, do it. You'll learn what problems the customers will have with the product, and make sure those problems are covered in your docs. (Ideally, all those problems will be fixed in the product itself, but in practice, the worst problems will be fixed and the rest will be renamed "features.")

Usability testing for the product and the docs should be performed with typical customers, not experienced developers or people who are more technical—or less technical—than your customer base. If you have video equipment, tape the sessions. Position the camera so it takes in the user, the written material and the thing being tested (keyboard and monitor for software, the unit itself for hardware). Don't answer any questions about the product, since you won't be there for the customer at their home or office.

Usability testing for the docs will rarely get funding, so it may be up to you to find a way to do it. The local least-busy administrative assistant is a often good person to test installation instructions and tutorials. Your mom or your kids might also be good candidates. But whoever you use, after a few times with similar products, they may become too experienced to represent typical customers.

You Are Not Your Own Best Tester

Face it. If you're able to write about technical material, you have a technical background and a natural feel for this stuff. Try as you will to avoid it, every so often you're bound to assume too much background or overlook a basic step along the way. Let someone who is representative of your audience read it through and work through the tutorials, if any. You'll be humbled and amazed. And your final product will be better.

Installation instructions, startup instructions, tutorials. Test them all. And test them again.

Gather a network of nontechnical people you can use as testers. Renew them over time, as they will become more and more technology-savvy from reading your work.

The Working Life

Employee or Contractor?

Most of this book applies to writers who work either as employees or contractors. There are differences and advantages and disadvantages to both ways of (writing) life. But both situations require the same skills and abilities, and hold you responsible for the same standards and duties.

Employee or contractor is a big decision—if it's yours to make. Depending on the industry, the economy or individual companies, you may have no choice.

Choice or not, there are pros and cons to both ways of working.

	Employee	Contractor
Pay	Steady paycheck, taxes taken care of. And benefits like medical coverage can be lifesavers.	Higher hourly pay, but higher taxes and more paperwork. Pay for your own medical, etc.
Stability	You know where you'll be working tomorrow and next month and next year (if you want to). Unless you get fired or laid off.	Unless you can line up multiple projects in advance—and none of them run late and mess up your schedule, you'll spend a lot of time between jobs digging up business.
Hours	Depending on the industry and your project's stage of development, you could work a straight 40 hours a week, or you could be expected to double that.	Feast or famine. You're either working 24 hours a day or looking for work.
Variety	Dependent on the company, you may have the opportunity to work on many different projects and teams, or you may sit alone in your cubicle and do the same thing day after day.	Get to work with all sorts of people and projects and companies. You also get to reject projects that you just don't want to do.
Responsibility	Dependent on the company; could be a little or a lot.	You better write a good bid or you'll be putting in a lot of free hours. If you blow a job and make a company miss a shipping deadline, then you may have trouble getting work in that industry again.
Advancement	If there's a writing group or department, you may be able to rise to the manager level, then on up the ladder, if you so desire.	As your reputation grows, you can command a higher and higher fee, but beyond a certain point, you'll hit a ceiling as an independent contractor.
Respect	If you're there day after day on project after project, you'll have the chance to earn the respect of your peers and bosses.	Widely varies. Some people will consider contractors "experts" and show respect and appreciate their work. Others will see them as hired stiffs that will be gone in a few weeks or months, and won't give you the time of day.
Politics	Company politics is a daily fact of life.	You'll still deal with politics on a regular basis, but at least you'll be able to trade in the jerks for other jerks.

	Employee	Contractor
Time off	Usually limited paid time off, but at least it's paid.	You can take off whenever you want for as long as you want, as long as you can afford it. Then again, not working for a long while can give competitors a chance to take over your best customers.
Socialization	You'll be dealing with a lot of people whether you want to or not. You'll have the chance to develop deep, long-term friendships.	You may be working in the customer's office or at your own place. You'll meet more people, but spend less time with them. Sometimes you don't talk to anyone for days at a time.
Attitude	You have the opportunity to ease back in a secure corporate environment, or push for new knowledge and experience. Of course, if it's a small company, even a startup, then you'll need a love of chaos.	You have to like change, and at least tolerate marketing yourself on a regular basis.
Ability to do a good job	You can work on improving the overall documentation system over a few projects, and lobby for more time and resources to do a good job.	You can use your good reputation (once earned) as leverage for the time and resources to do a good job.

Note: In some circumstances you might get a long-term contract, for months or even years, that puts you in pretty much the same position as an employee. Without sick pay, benefits, chance of promotion or other perks.

A preference? In my opinion, if the company situation is right, with a lot of variety and opportunities to learn and advance, then employeehood wins over contracting—at least for a while. Contracting, moving from job to job—possibly industry to industry—has a whole set of lessons to learn that an employee never learns. My suggestion is to do both for at least a while, then choose.

But when you really think about it, the difference between employee or contractor is all peripheral to the actual writing. It's the pay, politics and personal relationships that are different. The writing, the research, the customer's needs and your responsibilities to the customer and your work are the same.

Writing as Design Check

So you're working along on your manual, and you're getting near your deadline. You take stock, and pore through the document to see what's missing.

If you leave yourself **[more here]** or **[get this info from Larry, should be ready by July 3]** notes throughout the document, it is easy to find the missing bits and compile a list of everything you still need to do. Separate the list into those things you have the information to complete, and those you don't.

Items can make the list of don'ts because:

- You don't understand this button, process or control enough to explain it, or

- It's not in the product yet, so you can't test it, or

- Nobody seems to know how and when it will work.

If you organized your manual carefully and completely, then this list of the subjects you can't write about is also a list of all the design details that have been left for the last minute.

With your deadline looming, it's now time to track down some answers. Make an appointment with the appropriate person or people on the project; producer, designer, lead engineer or whoever is most likely to have the answers, and go through your list.

Also, choose who you meet with by disposition. Some people get rushed and cranky from end-of-project pressure, and might resent the time away from their main interest. Others appreciate both the need for good, complete documentation, and appreciate your list for the design check that it is. Anything and anyone that keeps details from falling between the cracks is good.

At this meeting, you'll hopefully find that most of the answers are ready for you; the design has been done, but the communication hasn't. Other things may be worked out and decided right then and there. (If you are allowed a voice in the discussions, push for the solution that is easiest to explain to the customer. This makes your job easier, but more importantly, it makes the customer's job easier, too.)

Most likely there will still be a few items that can't be decided without prototyping and testing. For these items, get a name and date to contact later for the final answer or decision.

When you leave the meeting, write up the results and distribute it to all involved, and a few others who would like to know. Include all the decisions made in full detail. Also include the unfinished items and the contact names and dates. Ask for comments and corrections.

Sometimes decisions aren't really decisions unless they're written down. People may say things or make quick decisions at meetings to get out of the meeting and back to "more important things." By writing it all down and distributing it to everyone at the meeting—and their boss—you make everyone take another close look at their decisions because they can be held to their word.

Never Take a Design Document at Face Value

Gather any design documents that you can as part of your research, but verify EVERYTHING with the actual product. Depending on the industry, the company, the development team and the individual in charge of the design docs, these documents could be very useful or total fiction.

Even the best will vary from the product. In those last rushes of building a product, the effort always gets shifted from the docs to the product, and the docs fall behind. Unless the person doing the design doc is also doing the manual, count on it being out of date.

If anyone hands you a really accurate, wonderful design doc, thank them, buy them lunch and don't tell any of your writer friends—try to work with those people again.

Maybe in the military development world the docs are kept strictly up to date, but I doubt it. In any event, no matter how good and accurate and up-to-date the design docs are, verify everything yourself with the actual product.

Development Speak—Alpha, Beta, etc.

If you have any exposure to today's technological world, you've heard the term Beta, or at least Beta testing. If you've ever worked for a technology company, you've no doubt run into the terms Alpha, Release Candidate and Gold Master as well.

What these terms actually mean within the development process is different at every company, and even varies from group to group within a company

Here's my best stab at what they should mean:

Alpha is the stage of the project when all the functions and features are in the product. They may not all work. There are no doubt many bugs to be fixed. But everything—every feature and function—that is going into the product is in the product.

Once Alpha is declared—often after an extensive product review by an outside board or team—only fixing, finishing and polishing should be allowed on the product. Nothing new should be added. No major changes should be made. (This corresponds to a feature freeze, explained in more detail in the very next section.)

Beta is the stage when, to the team's best knowledge, the product is finished and ready for manufacturing. At this time, the product is once again reviewed, then presented to Beta testers—both professional testers and typical customers—for punishment and abuse. The Beta tester's job is to put the product through the ringer and try to break it or find problems with it.

Release Candidate 1 status is reached after the product has been through a round of Beta testing, and all the problems found in that round are fixed. It is then sent out for another round of testing. If there are additional, new or unfixed old problems, then it goes back for debugging. The new version, Release Candidate 2, is once more put through the testing ringer. It is not uncommon to have a dozen or more release candidates, especially in software. When the product finally makes it through testing with no bugs or problems (at least none that the company's management can't ignore), then it's done. In software, this is known as the Gold Master. It's not gold, but it is the master that can be duplicated and shipped to customers.

Feature Freeze

Working on a development team with intelligent, creative people, who really care about their work and want to do the best job possible is one of the great joys of working life. It can also be one of the great frustrations.

There comes a point in any project, be it a book or a toy or a computer game or a VCR, when someone has to say, "That's enough! No more features, no more changes, no more nothin'! Let's finish this puppy and ship it!"

This is feature freeze: when you stop adding and changing, and just fix and finish. Feature freeze is something that writers come to love. After weeks or months of chasing a moving target (writing about features that change, disappear or suddenly invite their friends and relatives to join them), this is your chance to grab the target with both hands and throttle it. Or at least to catch up and finish your documentation.[30]

Depending on the development model, there may be multiple feature freezes in a project, allowing everyone to catch up and finish refining each stage of development.

Of course, once the final freeze (often associated with the Alpha stage of development) is in effect, the project isn't over. Not only do you, the writer, have to finish your work, but the rest of the team has to fix, debug, refine and finish the product. And that is usually a lot of work. A lot of boring, not-so-creative work.

The urge to keep adding features and trying to improve the ones already in the product is a powerful force that acts upon creative people. The post-freeze work for engineers, programmers, testers and others is drudgery compared to the more creative parts of the job. As the old adage says, "The last 10% of the project takes 90% of the effort." Because of this, everyone is tempted to break the freeze, and add just one more little feature. And some people, knowing that they shouldn't add things, will do it without telling anyone. Since each of those one-more-little-features has to be documented, or checked to see if it needs to be documented, you have to be ever-vigilant during the freeze, and hunt them out.

Feature freeze may be more important for writers in software than in hardware. Hardware, after it's totally designed and prototyped, still has to be reengineered for manufacturing, and that gives the writer time to catch up (or that may be when the writer joins the project). In software, the docs have to be written, laid out and printed by the time the last bug is fixed. One the program has passed Quality Assurance, it is ready to be mass-duplicated, packaged and shipped. If things keep changing during the final bug fixes, the docs won't be done on time—or they'll be wrong.

---

[30] *Writing is always catching up. You can't write about something until it exists. Well, you can, but you always have to verify it afterwards, and you usually have to rewrite or update it.*

If you're working at a smaller, less-organized company, there may not be a feature freeze. Ask for it. Push for it. Teach about it. Let the powers that be know that you need X amount of days after the product stops changing to get the document ready for layout.

There are a number of good books on software development that explain the importance of feature freeze. Ask the company to buy some copies for the project managers. Give them as presents, if you have to.

Feature freeze is good. Feature freeze will set you free.

Typical Writing Process Without an Enforced Feature Freeze

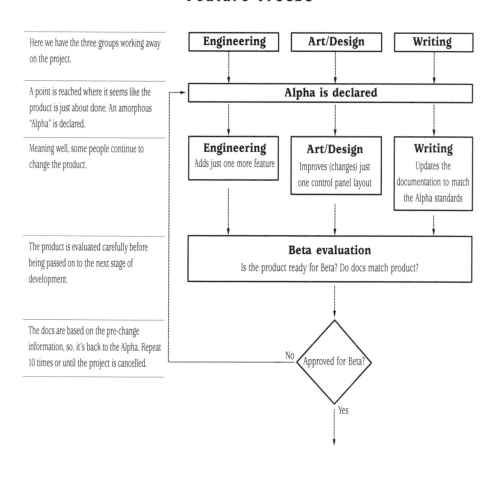

Here we have the three groups working away on the project.

| Engineering | Art/Design | Writing |

A point is reached where it seems like the product is just about done. An amorphous "Alpha" is declared.

Alpha is declared

Meaning well, some people continue to change the product.

| **Engineering** Adds just one more feature | **Art/Design** Improves (changes) just one control panel layout | **Writing** Updates the documentation to match the Alpha standards |

The product is evaluated carefully before being passed on to the next stage of development.

Beta evaluation
Is the product ready for Beta? Do docs match product?

The docs are based on the pre-change information, so, it's back to the Alpha. Repeat 10 times or until the project is cancelled.

No — Approved for Beta?

Yes

Typical Writing Process With an Enforced Feature Freeze

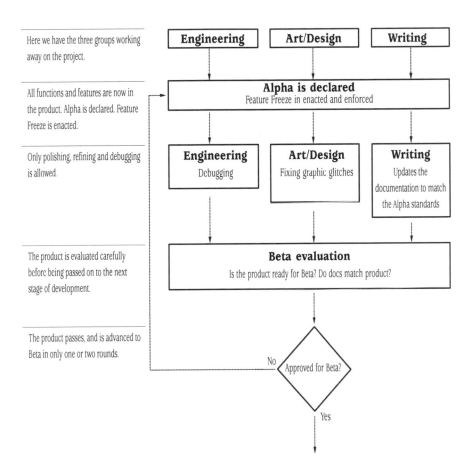

Here we have the three groups working away on the project.

All functions and features are now in the product. Alpha is declared. Feature Freeze is enacted.

Only polishing, refining and debugging is allowed.

The product is evaluated carefully before being passed on to the next stage of development.

The product passes, and is advanced to Beta in only one or two rounds.

| Engineering | Art/Design | Writing |

Alpha is declared
Feature Freeze in enacted and enforced

| **Engineering**
Debugging | **Art/Design**
Fixing graphic glitches | **Writing**
Updates the documentation to match the Alpha standards |

Beta evaluation
Is the product ready for Beta? Do docs match product?

Approved for Beta?

No

Yes

Resolution Meetings

When you have to meet your deadline, and your sources disagree on facts, the only way to get things resolved quickly and finally is through a meeting, sometimes called a design review or a resolution meeting. Email or memos won't do it. Get everyone in the room, including the highest-ranking project manager or producer you can, and bring up the issues, one-by-one, and calmly ask for the final, correct resolution.

Let them know that you will be writing down what they say, that you will run it by them once more after the meeting to confirm that you understood what they meant, and that they will be held to what they say. Whatever they promise, they are committed to deliver.

This will make them think things through and be careful about what they promise, but count on it changing again and again. That's just the nature of technology, especially software, and you'll just have to live with it. All you can do is minimize the hassle.

Often what you want to know can't be decided at that meeting, but you can at least get a tentative date when it can be decided and a name to contact on that date, then follow it all up.

Internationalization

The world is a shrinking place. Many US companies, even small ones, earn 50% or more of their money from sales in other countries. And this means that everything you write, as well as every label on the VCR, every bit of text on the screen, has to be translated into French, German, Spanish, Italian, Japanese and many other languages.

Once just called translation, *localization* became the term for translating and possibly repackaging products for various local (to them) foreign (to us) markets. In recent years, this process has been expanded to include major redesign of the product to fit the culture of various countries, and is now called *internationalization*.

Since internationalization is an international word, professionals in the industry can't agree whether it should be spelled with an "s" or a "z." What they can agree on is that it starts with an "i" and ends with an "n" and that there are 18 letters in between. That's why the common abbreviation of internationalization is I18N.

Good internationalization sometimes requires a major cultural conversion. Things we normally wouldn't think about can cause big problems. For instance, in some countries, the sight of a severed hand is incredibly repulsive. Makes sense. But this includes those cute little hand cursors we all have on our computers. Things like that have to be changed when you internationalize software for some countries.

I18N is a book unto itself (a number of them, actually—check out the Recommended Reading section), but basically, it involves timing, organization, and working with a number of other people: translators, project managers, engineers, designers, programmers, international Quality Assurance, foreign publishers and accountants, and on and on.

Much of I18N still depends on translators. For some jobs, you can hire almost any translator and get good enough results. Other times, such as for software—even simple software—you need someone with industry knowledge, who knows the standard words and phrases for menu items and dialog-box buttons in the country they are translating for.

Sometimes you may need someone very special, someone who will do far more than just swap the words from one language to another and clean up the grammar. You may need someone who will do a lot of rewriting to have the translated document cross cultural lines without losing its meaning—a true writer in another language.

There is currently a debate among I18N professionals about whether you can get a good translation from someone locally (here, in this country), or if it MUST be translated in the language's native country, where the native speaker hasn't been "contaminated" by even a few months in a different country.

Of course, even if you use a native speaker in their native country to translate, you'll still get complaints. Parisian French is different from Provincial French and German is spoken a little differently in northern and southern Germany. The safest bet is to ask your publisher or distributor in each country to recommend a translator, or see if they'll approve the one you want to use. Sound complicated? Welcome to international business and politics.

I18N and You

As a writer potentially involved with I18N, you need to be ready to work with a translator, locally or in foreign lands, to explain and reword any deeply cultural references that might not make sense in other countries. And there are more than you think.

When taking screenshots for a software manual, try to use shots that contain as little text as possible (if it still gets your point across), so they won't have to be recreated, reshot and replaced for each language. This might be tough in a word processor, but in a paint program or even a spreadsheet, with a little forethought, you can keep words to a minimum. Here's a hint: include the icon or button bar, but leave out the menu bar (words) whenever you can.

Ever wonder why software designers use so many icons on the screens, even when you can't tell what they are and it would be easier if they'd just use a single, clear word? It's because icons are semi-universal, while all the words will have to be changed in the program and manual for as many as a dozen languages. That's a lot of computer artist, programmer, writer and graphic artist work, and a headache for producers, project managers and asset trackers.

Besides, even if you have to change icons, they always stay the same size, while words in different languages vary. If you use words, you'll have to leave room in all the versions for the longest word in all the languages you're translating to (which can look bad with all those uneven spaces), or customize each screen of each language version (which takes a lot of programmer and tester time and is expensive).

This brings up the point that some languages take up more space than others. On the average, German takes approximately 30% more space than English. Individual words may be far longer than their English equivalents. And the German distributors I've worked with tend to not like abbreviations.

In your graphics list, keep track of which graphics may need to be replaced (almost any with words). You may be asked to supply all the screenshot versions for the different languages. There are two basic ways to do this:

1. Take new screenshots with each of the newly translated versions of the program, hopefully with someone who speaks the language sitting next to you. At the very least you'll need a page or three with a translation of all the menus, windows, etc.

2. Touch up the original screenshot, using a paint program to change the text to the various languages. Before you do, make sure you have a good, final translation that will match the actual program and the foreign language docs.

If you are sending files back and forth to someone in Europe, you may have trouble printing documents they send you, because your printer will be waiting for you to insert size A4 paper. A4, the standard size in Europe, is slightly smaller than our standard letter size. To avoid this problem, open the Page Setup dialog and change the paper size. [31]

Be careful with humor. I've encouraged the use of humor in this book, but when it comes to translations, humor has its own set of problems.

The best way to preserve humor is to have a translator with a sense of humor that can censor, change, rewrite or replace lines as needed. Translators like this are hard to find.

Other than that, play it safe and make sure that the humor can be cut without losing the necessary content. When you know that something is going to be translated, it is always best to go the safe route and stay away from anything that your mother would consider in bad taste. And, heartbreaking to some (including me) and joy-bringing to others, puns just don't fly after being translated.[32]

One last point. American English and British English are not the same language. Plan and budget for a translation between these two languages. (You may be able to ship an unchanged British product in the US, but not the other way around.) This translation is much easier and cheaper than, for instance, an English-to-French translation, but should be done. The main things you need to do are run a spelling check and have a native British-English speaker give it a careful once-over for slang.[33]

---

[31] *I've seen a whole office grind to a halt when the network printer stopped and waited for someone to change the paper. A dozen people had to scour the office, looking for the one who sent an A4-sized document to the printer and then left for a few hours.*

[32] *My first experience with puns and translators was on a software manual in 1989 or 1990. The translations, layouts and printing were done in the native countries, and we were sent the final product without a chance to comment. The French translated the puns literally, so they made absolutely no sense. The Germans left them in English (a reasonable solution). The Japanese chopped them out. There's a lesson to be learned here, but I'm not sure what it is.*

[33] *I've actually been punched by British friends when I said something innocent in American English that was rude and insulting in British English.*

Attend and Participate in Project Postmortems

A project postmortem is when the team members who have just completed a project get together and talk over what went right and what went wrong in the project. The goal is to identify those processes and techniques that worked and repeat them in the next project, and identify whatever didn't work and try not to repeat that. Depending on the company and the team members, a postmortem can be a casual meeting over coffee or a 200+ page report that is printed, bound and distributed throughout the company.

Even if you're a contractor, it's a good idea to attend the project postmortem. If nobody brings them up, throw out a question or two about the docs. Solicit suggestions for the next project, and use this information to lobby for more time and budget for the next project's docs.

The Style Guide

Many companies have a style guide. These are documents that define a consistent language usage for the company.

Besides some basic grammar, company conventions for document formatting, and commonly misspelled or misused words, the style guide will hold company- and industry-specific information, like proper spellings of commonly used industry terms, products or related companies, as well as lists of your company's trademarked and registered products.

If your company, or the company you contract for has a style guide, get a copy. Check to make sure that it is up-to-date. Look it over, and follow it as much as you can. It'll make the editor's job easier.

If your company doesn't have one, it may be a good job for a writer or editor during a slack period (if such a thing exists).

The Writing Team Handbook

If you are part of a writing or documentation group or department, or you regularly work with or subcontract with other writers, and you want your work to be fairly consistent, you may need a group handbook.

When you hire a new group member, bring on a temp or hire a subcontractor for a project, give them a copy of the handbook. It'll save you a lot of training time, and eliminate a lot of work.

The handbook may overlap some of the information in the style guide, but while the style guide is written for the entire company, the handbook should mostly have writing-specific information. Its main contents should include the group's duties, skills and procedures.

- The duties are all the different services that the group members can and will provide for clients or the customer. This is public information, both for the benefit of the group members and the clients. Post it or email it to people that may need your services.

- The skills list is actually a wish list of all the knowledge, skills and capabilities that you'd want your group to have mastery over. Every member of the team doesn't need every skill, but as many as possible should be covered by the whole group. Go over this list with new writers, and let them know which skills the group lacks. If they work on those skills the group lacks, they both help the group and increase their value to the company (good career move). Use this list when hiring new team members. Shore up your weaknesses.

- Procedures are the methods your group uses to get work done. They include the local writing/editing cycle, document maintenance, data backup, preparing text and graphics for layout, and any other standards and methods that work for your group.

Editing

"PROOF-READER, n. A malefactor who atones for making your writing nonsense by permitting the compositor to make it unintelligible."

— Ambrose Bierce, *The Devil's Dictionary, 1911*

"Editor: a person employed by a newspaper, whose business it is to separate the wheat from the chaff, and to see that the chaff is printed."

— Elbert Hubbard

This chapter covers the ways and means of dealing with editors and the editing process to fine-tune your writing and confirm its accuracy. This isn't an instruction manual for editors, but should help the writer to understand the editing process and to use this process to become a better writer.

The Basics

Red Ink Is Your Friend

If you're new to writing, get used to red ink.[34] And if you're not new to writing, stay used to it.

The first time I ever had my technical writing edited by a professional editor, I was shocked. I had been writing various unpublished efforts for what seemed like decades, and had written internal technical docs for an electronic design company for a couple of years. I thought I knew my craft. So, I wrote my first software manual for the consumer audience that would be professionally laid out and printed en masse. I had been over it a dozen times, honing and refining. The person I was contracting for really liked my work. Then he suggested we send it by an editor he had used before and was highly recommended by the graphic artist. I didn't think it would really be necessary, but what the heck. I was expecting a few corrections here or there, but when I got it back, there was so much red on it, that if that manual had been a living thing, it would have bled to death.

At first I was angry, then embarrassed and annoyed. Then I went through it carefully, and asked a few questions, and realized that there was a lot to learn here.

Over the next couple of years, every bit of writing I produced was sent by this editor before layout. And when it came time for me to hire a staff writer/editor, this guy got the job. And 10 years later, he's one of the people editing this book. (Say hi to the folks, Tom: "So many punctuation marks, so little time.")

Every time you see red ink on your work, it means that what you show to the world will be better and more accurate (as long as the editor knows what he's doing). That said, be sure to question your editors whenever you can and feel the need. They sometimes make mistakes and misinterpret your meanings (generally your fault for being un-

---

[34] *Editors generally use red pens to edit so the corrections stand out from the original writing. If two or more people edit the same manuscript, give them separate copies. If, for some reason they edit the same copy of the manuscript, they should use different-colored inks so it's easy to identify whose comments are whose, in case you have questions.*

clear in the first place). And when they're right, they give you a re-fresher course in grammar and teach you what an editor knows, what they look for, and what mistakes to avoid.

Red ink is your friend.

Editing Yourself

Unless you're one of those rare people who writes everything per-fectly the first time through (and if you are I don't want to know), you're going to be editing your own work, rewriting and reediting before you pass it on to someone else to edit. This isn't necessarily the most fun part of writing, so, find what works for you, and do it.

Some writers like to print out each draft, take the hard copy to a nice, quiet comfortable spot (or noisy coffee shop) and do their editing on paper. If you've got a decent printer, it's easier on the eyes than read-ing from a screen, and it gives you a chance to get away from the computer for a while. And there's something about holding a pen in your hand that's integrated with the thinking process, at least for those of us who didn't have computers at our desks in our formative years.

In spite of these advantages, I personally would rather read and edit my own writing on the computer screen than on paper. I hate writing more than three words in a row by hand. It's slow, it's sloppy (if you saw my writing, you'd consider sloppy a euphemism), and when you're done, you have to type it in eventually. So why not just do it on the computer in the first place?

Find what works for you, and do it. And do it again. And again.

> "Speak clearly, if you speak at all;
> Carve every word before you let it fall."
>
> **— Oliver Wendell Holmes**

The Joys of Being Edited

The experience of being edited is the old two-edged sword. It makes your writing better, but it hurts. It makes your work more effective, but that mean person said you were wrong. You learn more about your craft each time, but someone changes your art.

Well, get over it.

Finding and working with a good all-around editor is the best thing that can happen to your writing. If you're contracting and the client doesn't have a staff editor, find your own. There's bound to be an English major working as an administrative assistant or sales assistant that would love nothing more than to be able to put that degree to work. Find these people, and work with them. (And include their names in the credits.)

If there's nobody there, find someone on the outside. Pay them if you must, or find a way to barter (writing time for editing time) if you can.

Relationships With Editors

If you have regular editors you work with, you need to cultivate a good relationship with them. When possible, have editors go over their edits with you, both for your own learning (in the instances when they're right), and to make sure that they understood what you meant in those places where you might have been unclear.

If you hand things off and they make their changes and you never see the document again until it's printed, try to change the process. You need the feedback to grow in your writing.

The more you work with an editor, the more that person gets to know your voice and your style, the less they'll mess with your stuff, and the better you'll look.[35]

If you're a contractor, develop a good relationship with an independent editor or two, and the two (or three) of you can help each other find jobs. Networking is good.

> "There is probably no hell for authors in the next world—they suffer so much from critics and publishers in this."
>
> **— C . N. Bovee**

---

[35] *Here's an instance where pluralizing* editor *to use the pronoun* they *would take away from the point of getting to know a particular editor, so I resorted to using* they *as a singular pronoun.*

What's in a Name?

Editors go by many names: editor, associate editor, line editor, copy editor, production editor, proofreader, technical editor, developmental editor, editor-in-chief, and sometimes even lead writer, director of creative services or V.P. technical communications.

Various editor titles mean very different things at different companies. Wherever you are, learn what the titles mean and who does what.

This is a book for writers, so rather than digging deeply into editing jobs and titles,[36] I'll cover the types and stages of edits that your work should pass through, and ignore the job titles.

---

[36] *For a good description of how the editing process works at a very large, very established company, see Judith A. Tarutz's book,* Technical Editing, The Practical Guide for Editors and Writers.

Process

Types and Stages of Editing

These stages have different names at different places, so find out the local terminology and use it.

Structural Edit (Outline Edit)

This edit is generally performed by your boss, who could be some variety of editor or head of the writing group. It is generally done after the initial research and information gathering, and you've put together your first detailed outline of the docs. This edit looks for completeness of intended information, subject areas that are too far off the mark or inappropriate for the audience or document size, proper organization of subjects, and conformance with the company's document organization style.

Requested changes here are at the section, chapter or whole document level.

First Draft Edit (Content Edit)

This is the biggest, and generally most stressful edit for the writer, because it gives feedback on the accuracy and completeness of the actual content and the writing style and quality. At this point the main editor may suggest (or demand) changes that could include rearranging information and even cutting or adding paragraphs.

As always, different companies do this in different ways, but I find it best to print up copies and distribute them to:

- Your boss (who probably did the structural edit)
- Any other editor your boss chooses
- Your peers (other writers), if you so choose
- Various people on the project who should be able to check your content for accuracy, including engineers, producers and testers
- Others around the company, including the marketing product manager and possibly someone in senior management, who may not

give it a lot of attention, but wants to be kept in the loop and needs to be assured that progress is being made

This can be 10 or more copies, which may seem like a lot, but you'll be surprised how each person will catch something different.

Make sure all the distributed copies have the editor's or reviewer's name on it, so you can ask the right person questions when you don't understand their comments or can't read their writing, and so you can weigh conflicting comments. Gather all the edited manuscripts and condense all the edits and changes into one master, then input your changes to the document from that one.

If there are conflicts between edits—and there will be—go to the proper source for resolution: lead engineer and project manager/producer for content, and the appropriate editor for grammar and style.

Depending on the size of the project, its level of organization and the responsiveness of the team, you may have to corner a smaller group of engineers, producers and testers for a second or even a third content check.

> "Only ambitious nonentities and hearty mediocrities exhibit their rough drafts. It's like passing around samples of one's sputum."
>
> **— Vladimir Nabokov**

Corrected Manuscript (Line Edit)

This edit, possibly performed by the same person as the structural edit, possibly by someone else, is done after the changes required by the first draft are done, and you're sure that the content is as accurate as possible. Line editors should check for grammar, punctuation, typos, clarity, reading level and general conformance to house style. The changes here should be at the sentence level. Hopefully, no major structural changes, additions or deletions will be required at this time.

Depending on the writer and the job, this edit may be performed more than once before the document is passed on to layout. The last time it is done (even if it's also the first), the document should be as complete as possible, including graphics (as separate files, not necessarily in the document), title pages, graphics list, table of contents and style sheet. If you're performing the final layout or graphic arts will be laying it out in the same program that you used to write it, then include an index, too. If someone in graphic arts will be importing it into another program for layout, then skip the index until the layout's done—indexes generally don't export and import well.

Layout or Page Proofs (Visual Edit)

This edit is primarily concerned with the work done by the graphic artist. The editor(s) generally won't read every word, but as they scan, if they find something that should be changed, they'll flag it. There shouldn't be much. At this point, making changes is very expensive. Adding a word or two can cause many pages to lose alignment and require going back to layout for hours.

Typical things to look for in this edit are: general visual consistency and conformance with company guidelines, line spacing, alignment of text, widows and orphans, pagination, headers and footers, matching page numbers in table of contents, index and cross references, accurate and up-to-date legal/copyright info, proper graphics in the proper place, captions matching the graphics, and the general look and feel of the page.

Before this "official" edit, you, as the writer, should be in contact with the graphic artist, making sure the document imported completely (if the doc is converted to .RTF as part of the process, tables may disappear), that all the graphics are there, and that the artist understands all your notes and directions for captions and callouts. You may go back and forth a number of times before the theoretically complete document goes to the visual edit.

Blueline

As printing gets more and more high-tech, I see fewer and fewer bluelines. But if there is a blueline in the printing process, give it a final, careful check. You definitely don't want to find grammatical or content errors now. If you do—and they're too bad to ship—printing will be delayed and a lot of money will be wasted. Don't be the one responsible for this—make sure the document is edited very carefully before blueline.

The things to look for in a blueline are the overall look, missing parts of letters, pagination, clear graphics and correct text flow from page to page. This may be done entirely by an editor and not involve you, but you should know if and when it happens.

The Editing Funnel

The range of what editors should look for—and should find—changes throughout the writing project, starting very wide, and narrowing with time.

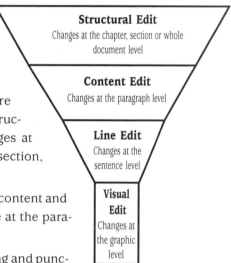

In the earliest stages, editors are looking at the big picture; overall structure and organization. The changes at these early stages will be at the section, chapter or whole document level.

The next round of editing looks at content and style. Changes at this stage will be at the paragraph level.

Next, the editor will look for spelling and punctuation and sentence structure. Changes are at the sentence level.

Finally, the edits are concerned with the visual aspect, which is more in the domain of the graphic artist than the writer.

Why narrow the focus at each stage of editing? Two reasons:

- **Money**—the later in the project that you make changes, the more it costs. Not only is there extra editing time if the editor looks for everything every time through the document, but there is more writer time, then more editor time to check the writer's changes. If a document has structural changes after it has been laid out, then you throw away a lot of work and money, and basically go back to the drawing board, rewriting, reediting, re-laying out. If you make major changes at the blueline stage, you're wasting even more money and blowing the printing and manufacturing schedule, which could delay product shipping. And if you need to make changes after the document has been printed, somebody's job is in deep trouble.

- **Getting it done**—you have to finish the project sometime, preferably on schedule. You've got to keep the project moving forward. Do a good job at each stage, do a good edit and correction at each stage, and move on.

Of course, if you find problems at late stages, you should fix them if you can. To avoid the need for late, expensive rework, each edit at each stage must be complete. Any rework that is done must be properly written and rechecked.

Sending Docs Out for Editing

For a structural edit or a final line edit that is done by one or two people who are as aware of the schedule as you are, you don't have to worry too much about the edit getting done on time and the document being routed to the right place. But for the first-draft content edit, where there are lots of copies floating around, and the reviewers aren't professional writers or editors and have many other things on their minds, you have to worry.

Attach a cover letter to each copy you distribute for editing and review. On this letter, include:

1. Reviewer's/Editor's name

2. Date and time it is due

3. Where it should be delivered, or who to call for pickup

4. What they should be looking for (general content accuracy, grammar, or just the accuracy of their particular area of expertise)

5. Special notes to the individual: i.e., "You're welcome to look over the whole thing, and comment on everything, including the grammar, but if time is tight, what I really need is for you to check the accuracy of the Inside the Simulation section."

6. State that many people are editing and reviewing the document, and every comment will be read and appreciated, but not every suggestion will end up in the final document

7. Thank them in advance

8. Your name, your phone number and extension, your email, your location in the office in case they need to find you

9. Also put the reviewer's or editor's names on each copy as well as the cover letter, in case the cover letter gets lost

Other things that help with this edit are:

- If you can, deliver the document in person, introduce yourself and explain what you need from them.

- Keep a checklist of those you've given the document to.

- Send out a reminder, by email or in person, enough ahead of the due date so they can actually get it done. Try not to seem like you're nagging while you're nagging. (This may be a good time for bribery. If you really need one particular person to check the content of a section, chocolate or a good cup of coffee can work wonders.)

- For any documents you send out for editing, make sure there are page numbers. You'll need them for locating, comparing, discussing and compiling the information in the drafts. Yes, the page-number information will be lost when the document is imported into a page-layout program. But put page numbers on the docs you send out for editing.

Reviewing the Reviewers—and the Design

When you send out multiple copies of a document for review, you not only learn a lot about your document and your writing, you also learn about the various individuals who edited your work and the general company attitude about documentation.

Use this learning experience to find out who are the most valuable reviewers and editors, and who you shouldn't bother—or bother with—next time.

First and foremost to learn is who actually comes through and delivers on time.

Beyond that, look at the quality of the edit or review. You'll find good feedback in places you wouldn't expect. I know a programmer who is the biggest stickler for grammar and has an amazingly sharp eye for finding mistakes. And there are some people who just have to leave their mark on everything by making some sort of change or negative comment.

> "Nothing, not love, not greed, not passion or hatred, is stronger than a writer's need to change another writer's copy."
>
> — **Arthur Evans**

You'll be amazed by how many people will change the same sentence or fact in completely different ways. When this happens, look closely to see if you caused the confusion by being unclear in your writing, if it's just a difference of opinion, or if it's a problem with the design.

While you're reviewing the reviewers, you can also review the completeness of the product design—or at least the completeness of the communication of the design. Unresolved design issues and those design issues that aren't properly communicated to the whole team will jump out at you when two or more reviewers disagree on the content (let them disagree on your writing or grammar; you can resolve that with an editor). When this type of conflict arises, it's time to talk to the person in charge of the project. If necessary, they will call a resolution meeting. Be careful how you handle major design or communication conflicts. Depending on how you handle them, you can come to be known as a valuable team member who found a potential problem in time to solve it before it caused damage—or a nitpicky jerk. Be positive and helpful. This is part of your contribution to the product above and beyond the documentation.

Never Underestimate the Value of a Fresh Set of Eyes

No matter how many times you—or your editor—look something over, you're bound to miss something. And looking again and again may not find it, because you're seeing what you expect to see.

I don't have the psycho-perceptual research to back this up (if you know, please fill me in), but I've seen it dozens of times. You think you've got something perfect and hand it over to your editor, smug in your perfection, and the editor finds many mistakes that you can't believe you missed. (You missed them because you saw what you expected to see. Once you've looked at something a few times, you just can't look at it objectively enough to see what's really there.) Then when the editor is done with it, a last-minute tester will find more errors. (The same thing happens to the editor, once they've been over it a few times: they see what they thought they saw. See?)

Most books I read, fiction or nonfiction, contain typos, errors or confusing sentences. I see them because I'm approaching the material with a fresh set of eyes.

Bottom line: No matter how careful you and your editor are, there are bound to be at least some errors, typos or missed facts in some if not most of your documents. The best way to minimize them is to have another person or two or three look it over with fresh eyes.

Peer Reviews

When I was managing the tight-knit writing group at Maxis, the writers started to run their work by each other before each edit (which meant before I saw it). This worked so well that it became a regular practice.

It really was one of those win-win-win situations:

- The writers got good feedback from peers in a nonthreatening situation. The reviewers were always polite and helpful, never rude or vicious, because they knew their turn would be coming up soon. Because the drafts were a little more polished, a little more thought out, with the "controversial" points discussed in advance, the writer presented a better draft to the boss or editor, and, as a result, looked better.

- The reviewers got practice in editing and analyzing documents, which made their writing better.

- The editor was presented a more polished product, so the editing went faster. (This was particularly important to us because we had no full-time editors. Those who edited—and those who managed the group—had their own writing projects to finish.)

People and Politics

"Crime does not pay ... as well as politics."

— Alfred E. Newman

"The word 'politics' is derived from the word 'poly,' meaning 'many,' and the word 'ticks,' meaning 'blood-sucking parasites'."

— Larry Hardiman

This chapter covers anything to do with people and companies. Some of it has to do with the regular course of your job, but much of it has to do with company and office politics.

On the Job

Job Titles

I began writing about technology in the early 1980s, while I was working at a small communications electronics company. Because it was only part of my job, and considered a very minor part, writing wasn't involved with my job title.

Later, when I began working for an even smaller entertainment software startup, writing was a bigger part of my job, but, being a wacky bunch of creative young people, we chose job titles to express our individuality, not to communicate our job functions. A couple of the more conservative titles at that time were King of Customer Service and Human Resourceress. For the first couple of years, my job duties were all over the place, ranging from writing to testing to customer service to tech support to various management tasks—a typical startup. My job titles changed when it was convenient (whenever it was time to print up a new set of cards).

Then, as the company grew and staffed up it became time to specialize, to pick a job or job path. And that's when we started to establish official-sounding job titles.

The path I chose was writing. It was what I enjoyed the most. The title took a little thinking. The biggest writing tasks were writing manuals, which immediately suggests the title Technical Writer. But that didn't really cover the job. We (there were two writers by this time) also wrote press releases, marketing materials (box copy, sell sheets, etc.), magazine ads, all sorts of internal policies and other documents. And, of course, we did our best to add a little flavor, color and humor to the games we made. The title Technical Writer, respectable though it is, just didn't cover the type of job I had, or wanted to do. I felt it would be limiting. Names and titles are boxes. I believed that if I was thought of as a technical writer, then I'd end up only writing technical stuff.

After going around and around, we finally settled on the title Creative Writer. We would have Creative Writers, Senior Creative Writers and the Director of the Creative Writing Group. Yeah, it sounds somewhere between pretentious and dorky, but it was the best we could come up with at the time. Dorky or not, it did allow us, for many years, to write, rewrite, edit and influence all sorts of documents, technical and otherwise. It was a good run.

What's the point? In the context of this People and Politics section, the point is that people's perceptions and opinions of you as well as of your work can be biased by titles and preconceptions. If you deal with people—and it's unavoidable—be prepared to face and battle preconceptions. Watch for boxes that limit you and your potential. Don't let job titles or the expectations of others limit your creativity.[37]

But for writing and communication in general, the point is that we should produce better, more readable, less intimidating, totally useful documentation. Then maybe the title Technical Writer will be one of respect outside of our own departments and societies.

"Bad work follers ye ez long's ye live."

— James Russell Lowell

A Job by Any Other Name...

Many of the things in this book that I propose writers know and do may not come under the job definition of "writer" in your company. While at Maxis, I did my best to expand the definition and duties of the writer's job.

Product design and development is a combination of creativity and analysis. But it depends on communication. Communication above for project approval and funding, communication below to keep everyone on the same track. My thought was to let the creative yet analytical people with a skill at communicating—writers—use their talents for the good of the product, the company and themselves. It made business sense, and it was more challenging and more fun for the writers.

A writer was assigned to each project as it started up. Depending on the project needs and the skills of the other team members, the writ-

---

[37] *Maybe there are technical writers who don't consider themselves creative. Who don't have a novel or screenplay in the works or on the back burner just for the joy of writing and creating. But I haven't met any.*

ers may have had one, two or more projects at a time. They were involved with the design document, the interface design, research, all onscreen text, user-testing (of the product as well as the docs), bug testing, marketing materials, sell sheets and package text. And, of course, the manual and other documents, both printed and onscreen.

It took a while to establish this wide-ranging role, but once the value of it was proven, producers kept budgeting more and more writer time, until, eventually, most projects had a full-time writer. Near the end of the project we would often add additional writers to get the manual and online help done on schedule, but the original project writer was on it from the beginning. This writer, the lead writer on the project, knew the project through and through and managed and coordinated the additional end-of-project writers (which made my job easier).

After a number of years, the company was bought out by a much bigger company with their own way of doing things. When I explained what my group did, I was told that a lot of what we did wasn't a writer's job. I was told that writers write the little booklets that come in the CD jewel boxes, and occasionally, marketing materials. All the other tasks we performed were necessary, but they were done by people with other job titles. When the takeover smoke cleared, most of the writers were given different job titles, like associate producer or game designer. But at heart, everyone still thought of themselves as writers.

If you think of yourself as a writer, I encourage you to push to get out of your box. Learn new things, take on more responsibility and use your creativity—it'll make you happier in your job and make you a better writer. If it takes you to a different job title, so be it. Use your skills, expand your capabilities and have fun. That's an order.

If you don't think of yourself as a writer, I say, "Writing's not just for writers any more!" If you're a producer, designer, engineer, project manager, product manager or in almost any management job, you need to use written communication to get your work done. While you're writing, you are a writer. Even if it's only for a few hours a week, it's worth doing well.

Newer and Smaller Companies

Working at a newer or smaller company has both advantages and disadvantages.

The advantages are flexibility, excitement and the ability to wear many hats. The disadvantages are flexibility, excitement and the necessity of wearing many hats.

Newer and smaller companies are generally more chaotic, which means more fun for those who like chaos, and torture for those who like order. If you're an employee (as opposed to a contractor with a contract limiting your duties to a specific purpose), you will be expected to wear many hats and do much work that isn't in your job description (if you even have a job description).

On the other hand, larger, more established companies have their advantages and disadvantages as well. The advantages are that everything is defined, and it is clear what you do and don't do. And those are the disadvantages, too.

Before you take a staff position at a company, make sure it suits your temperament.

> "The dictionary is the only place where success comes before work."
>
> **— Arthur Brisbane**

Chaos for Fun and Profit

Some ... OK, a lot, of what I've written may give the impression that product development—especially software development—is an unorganized, chaotic mess. At times it is. At times, it should be.

Projects can be located on a scale from total chaos at one end to total regimented control on the other. Designers generally like things to be closer to chaos, managers like it closer to regimented. Neither is right or wrong. It depends on your goals. Designers want and need creative time and freedom. They need to be able to change directions, follow hunches and explore possibilities. Managers are paid (and therefore need) to get the job done on time, on schedule and on budget.

So the goal is controlled chaos. Enough randomness and spontaneity to keep things exciting and creative, but enough control to keep the project under control and actually finish it and ship it. A well-run project is a compromise, somewhere in the middle of the scale, starting closer to chaos, and slowly moving toward control as the development progresses. Where on the scale you start and stop depends on the type of product you are making.

Yes, creativity (and therefore chaos) is necessary in creating any product, no matter how serious or businesslike. But there is a difference between creating a business application, like a word processor, and a game or other entertainment software.

For a word processor, you can look at your last product, look at the competition, poll the customer base and come up with a feature set that you know will make a viable product. Creativity helps with the design aspects and special features all along the way, but it is primarily a project of implementation. Games and entertainment, on the other hand, have no guarantees up front. It's entertainment. Nobody really knows what will be a success. Look at movies. If people knew what made successful entertainment, there wouldn't be so many multimillion-dollar failures. That's why with games it is important to keep things a little more chaotic, to allow the creativity to flow as long as possible before feature freeze.

Games may seem frivolous to some, but they may be the hardest software to pull off successfully. And to add to the difficulty, not only do they require more chaos and luck, but they also generally require more technology. Games are pushing the limits of computers (and helping to sell the hardware), with their extensive use of graphics—2-D and 3-D—sound and processing power, and, of course, documentation.

Getting back to the subject, my final summation on all this is: chaos is good, up to a point. Learn to use it, learn to live with it, learn how and when to control it, and above all, learn how much you can take, and choose your jobs accordingly.

Business and Politics I for UnTechnical Writers

It is very likely that you, as a writer, will sometimes be prevented from doing a good job. There won't be the budget, the support, the time or even the general desire to provide the customer with enough information to use and enjoy the product. You want to do a good job. You want the customer to feel good about your work and the product. But the local attitude doesn't care.

It's almost impossible to change people's attitudes, but it may be possible to change their minds.

The way to get the support, time, space and money to do the job right now is to show how it will cost the company less in the long run. Convincing professional business people to spend more money up front takes some extra work, some stretching beyond your strict writing duties, and the gaining of an understanding of both business and company politics.

Some people, especially if they're new to writing, believe that writing is above or beyond all that business and office politics stuff. Get over it. Writers may spend a lot of time alone, but at some point—if they want to get read or paid—they have to deal with people. If you deal with people, you deal with politics. If you deal with money, you deal with business. And in the technical world where even computer games can have multimillion-dollar budgets, you're dealing with a lot of money.

I'll go into this in detail in the next section. For now just take a little time and absorb the fact that like it or not, business and politics will play a part in your writing life.

> "Politics is the science of how who gets what, when and why."
>
> **— Sidney Hillman**

Business and Politics II for UnTechnical Writers: Using the Bottom Line

Sometimes, in order to provide the best product documentation that you can, you've got to do more than write. You've got to fight for the resources, pages and Cost of Goods to do the job right.

Resources are people, space and equipment. That's hourly wages or salary plus benefits for writers, graphic artists and anyone else who will spend time contributing to, editing or checking the docs; rent on the office space; plus computers, printers, software, copiers, etc. A lot of these costs "shouldn't" count, since most of those people and copiers would be there anyway and cost just as much whether or not you add work, but this is the kind of thing that senior management thinks about, so be ready to deal with it.

Pages, as used here, means being allowed enough manual pages and other pages, like Quick-Start cards, to explain everything that needs to be explained.

Cost of Goods (COG) is how much the product costs to manufacture (including printing, duplication and assembly, but not research or development). Adding a Quick-Start card to a box of software may cost .02 in materials, .03 for printing, and .05 to have someone drop it into the box during assembly. That's a dime. Not a big deal? Those dimes add up when you manufacture hundreds of thousands or even millions of units. If you do go to the higher quantities, a good manufacturing manager can beat the price per unit down, which helps, but it still adds up to a lot of money that may need to be justified.

Here's an example:

Let's say you're just about done with a software project. During your final testing, you come to the conclusion that many customers will have difficulty installing it (a very common, very costly situation that should never happen, but does). Sure there are complete instructions in the manual, but from tests (putting a mock-up of the package with all the materials into the hands of some typical customers) or from a reliable source, like your customer service or technical support manager, you find that your typical customer won't dig into a big book just to get the program started. Many will experience frustration, and either quit or call for tech support before opening that manual. Those that do open the manual will resent it.

Your first and best solution is to have the install program reworked to be simpler, with more onscreen explanatory text. You talk to your project manager (or producer or product manager or whoever's in charge), and find that it's too late to change the program. Any changes will require an additional round of testing and QA approval, which will delay the ship date by a week. Besides, someone already sent the master off for duplication.[38] A change now would require throwing away thousands of freshly pressed CDs. She[39] asks you for another solution.

You believe that putting a Quick-Start card in the box with simple installation instructions will set things straight. It can be the first thing the customer sees after opening the box, and brightly colored, so it can't be missed. It'll guide the customer through the process, avoiding all the snags that your test group encountered. And it'll be a simple, single-page, one-sided (cheaper than printing on two sides) card with no folds (which add to the cost).

You explain the new solution to your project manager, and she likes it. But an hour later she calls and says that it's not in the budget, and her request for a higher COG has been turned down.

Now it's decision time. Do you give up and move on to the next project, knowing you tried? Or do you try to find a way to do the job as "right" as you can? It's your decision. Depending on your position there (employee or contractor), the local attitude about documentation (vital part of product or necessary fluff), the company's attitude about the budget (tool for planning and controlling costs or the ultimate, unassailable word from on high), and the people you work with, the best decision may be to walk away. On the other hand, if you think you have a good chance of reversing the decision and getting that card into the product, you may want to try.

Of course, you can't march into the president's office and make demands. Well, you can, but it won't help. You've got to be prepared to

---

[38] *Yes, it does seem odd (or wrong) that the program would have passed QA and gone off to manufacturing without testing the installation process with typical customers. But the reality of the software world is that many companies still consider the install program a last-minute slap-on task—as opposed to the first customer impression of their new product—and try to get it done as quickly and cheaply as possible.*

[39] *Pronouns again. Here is a situation where pluralizing to avoid using either* he *or* she *didn't work, since it is about a single individual. Using* they *just didn't sound right. Sometimes it does, sometimes it doesn't. I didn't want to use* he or she *because it also sounded wrong. In this context, it was a person the writer and reader would know on a personal basis, and the wishy-washy either/or approach didn't feel right. So I flipped a coin.*

Just In Time Manufacturing and You

The current popularity of Just In Time (JIT) manufacturing means that companies no longer manufacture and stockpile many months' worth of product, but only make the product or parts of the product as needed. This keeps the cash in the company's coffers (and banks and investments) longer, and cuts down the cost of the space to store the finished product and parts until it's time to ship it to the stores. These savings are weighed against quantity discounts (up to a point, the more of something you manufacture, the less it costs) to set the manufacturing cycle.

The good side of this is that it can give you more chances to update the product, if necessary. The bad side is that since the product can be updated, some companies will ship on schedule, whether or not the product is really ready, figuring they can fix it in the next manufacturing cycle.

Also, some parts are more just-in-timely than others. For instance, while large duplicators can crank out a hundred thousand CDs overnight, a print run of the same number of 100+ page manuals will take two or three weeks. Because of this extra time, manufacturing managers are likely to stockpile a lot more manuals than CDs, meaning that manuals can be updated less often.

justify the increased expense. How? There's only one way: prove that it will save money in the long run. If it won't save money in the long run, then don't bother fighting. Instead, work with the project manager to see that the installer will be updated by the next manufacturing run.

How do you do this?

1. Think like a businessperson.
2. Gain allies.
3. Gather information.
4. Put together and present a plan.
5. Be prepared for a grilling.
6. Accept the judgement.

Thinking like a businessperson—Even though the presence of that card in the box may be a matter of personal pride, or even a moral issue for you, the decision will be based on business principles, with a sprinkling of politics. Approach the issue from the business perspective.

Gaining allies—Sometimes referred to as *consensus*, having others who agree and believe in your issue is of the utmost importance. This is the politics. If you can't convince your peers in various departments and your immediate superior or two, you won't get very far with the people who OK budget changes. First and foremost among allies is your project manager or producer. If she's not willing to push for this, then give it up. But if you can make a reasonable case that what you want will make the product better without costing too much or delaying shipment, chances are good you'll get support here. Beyond the project manager, look for others in the company who will personally and professionally benefit from your plan. You'll find these people in customer service, tech support and marketing.

Gathering information—This means real numbers. Numbers from customer service and technical support, like:

- The complete cost to the company of handling each customer service and tech support call (including salaries, cost of calls, cost of space, cost of equipment, cost of administration).

- The typical number of customer service and technical support calls that are received during the initial release of a product.

- The projected number of these calls if the product ships without the card.

- The projected number and cost of temps that will be needed to handle the extra load during the release.

- The projected number of these calls (hopefully smaller) if the product ships with the card.

Numbers from marketing, like:

- Percentage of focus group that had problems with installation.

- Estimated impact on company image.

- Estimated impact on sales of upgrades to frustrated customers.

Numbers from manufacturing, like:

- Change in COG.

- Time impact on shipping.

- Time impact on manufacturing planning.

Putting together and presenting a plan—If your numbers justify adding the card, then start putting together the plan. If not, then abandon it.

In addition to long-range cost savings, your plan should include all times and costs needed to complete and add the card. Don't leave any costs out or try to slip any by. You'll either be blasted out of the water right away, or it will come back to haunt you when the real costs are tallied, and again the next time you push for something.

If the proposed card doesn't delay shipping (big, big issue), point it out. Your plan should also include the steps to update the install program before the next manufacturing run so you can eliminate the extra card (extra cost) as soon as possible.

When it comes time to present the plan, let the project manager do the talking—even if you've done all the prep-work. It's her job to do it, since she's generally responsible for the product's financial success. Be there as a resource.

Preparing for a grilling—The powers that be may really be looking for a hole in your plan, or they may just want to make sure you've done your homework. Have answers and backup data ready.

Accepting the judgement—If all goes well, then great. Get to work. But if you have the numbers that justify your cause, and you're still turned down, walk away. There may be other political issues involved, like the head of product development may not want to admit that he signed off on a badly developed installer, or the director of customer service may want an inrush of calls to help justify a higher budget (and a bigger empire). It is better politically to be known as somebody who fights for what they believe is in the best interest of the company, but knows when to quit, than to bang your head against a brick wall and get a reputation as a troublemaker. Believe me. I have the lumps to prove it.

"If you work for a man, in heaven's name work for him! If he pays you wages that supply you your bread and butter, work for him—speak well of him, think well of him, stand by him and stand by the institution he represents."

— **Elbert Hubbard, Get Out or Get in Line**

"Work, it seemed to me even at the threshold of life, is an activity reserved for the dullard. It is the very opposite of creation, which is play, and which just because it has no raison d'être other than itself is the supreme motivating power in life. Has any one ever said that God created the universe in order to provide work for Himself?"

— **Henry Miller, The Creative Life**

Oh, the Humanity

Bribery and Niceness

Office politics. If you've ever worked in an office, you know that maneuvering in the political landscape can be as important to getting your project finished as the duties in your job description.

Sometimes when you gotta get things done quickly in an organization, you gotta grease the wheels, or at least bring yourself and your project to the attention—in a good way—of those whose help you need.

Maybe handing out hundred dollar bills would do the trick ... I've never tried it. But dropping off a service request to the assigned graphic artist and using a nice piece of quality chocolate or a latte as a paper weight—just so it wouldn't get blown off their desk by the ventilation system, honest—always seems to get the job moved up in priority.

The trick here is knowing what local customs allow (don't get into too much trouble), knowing what legal, mostly safe and non-impairing treats your co-workers like (don't give them anything that might impede work getting done, ruin their diet—too much—or thwart their attempt to give up caffeine), know who this will work with and who won't respond, and don't expect someone who you normally don't even talk to to respond well. If it seems like a bribe, then it won't work. This is where niceness comes in.

You can't expect to be best friends with everyone in the company, but it pays to have at least a good rapport with the people you need to work with. Be a little extra forgiving for those whose work, opinions and reaction times affect your ability to get your work done. I'm not suggesting that you kowtow to jerks, but be understanding. Most of the people you work with are under as much pressure as you, and are likely to blow off a little steam now and then. Reacting too much and too negatively to a first sign of crankiness can set you up as a permanent target. Stay out of the line of fire; let it pass. And if you know someone is stressed out and up against deadlines and doesn't have time to respond to your request, don't yell and scream and demand. You'll just be shooting yourself in the foot. If you really need something from them, then you might be able to offer a trade; take some other task off their hands to free them up for your project.

And then again, there's the totally legitimate method of bribery. Depending on the company, you may be able to officially requisition funds to help you get some things done.

If you need people to read and edit, or test something like a tutorial, call a brown bag lunch and have the company supply cookies and sodas, or have them spring for the full lunch. The fact that it's on a lunch hour saves the company a lot of money.

There are other ways to use a small amount of funds to get big results. Use your imagination. But try to stay legal.

Control Issues

As always, there are control issues.

Who has the final say on wording, jokes, layout, etc.? The writer? The producer? The editor? The head of the writing department? The president of the company? This can work (or not) in a lot of different ways, depending on the circumstances, company, industry and individuals involved.

On any job you have or contract you take, find out who has the control—who you are really working for—and what they want to see.

If the designer or lead programmer wants you to put in lots of humor and sarcasm, and you do, then the guy who's really in charge makes you rip everything out and start over, you've all wasted a lot of time and energy.

Never Take Any Information for Granted

If I had a dime for every time a developer or producer said, "This is the way this feature's going to be, so go ahead and write it up like that and send it off to layout on schedule," I'd be a rich man.

If I had to give back two dimes for every time the feature actually ended up working the way I was told, I'd still be a rich man.

Documentation rule number something or other:

It ain't over 'til it's over.

The way schedules (especially in software) work (or don't work), you may be expected to finish the documentation, shepherd it through layout, and send it off to the printers before the product is actually finished. This happens more often in smaller, newer companies, but it can happen anywhere. I've seen tens of thousands of manuals printed on schedule, based on the sworn promises from every producer and programmer on the project that all the information in those manuals was completely accurate and that the final program will appear in two weeks, perfectly matching their sworn testimony of accuracy.

Four to six months later, when the program was finally in its last testing stages, it bore little resemblance to the manuals, which were now more accurately termed landfill (or recyclable material).

Of course, by this time, the writers and editors and graphic artists had revised the manual a few times, and an accurate version was ready by the time manufacturing began. But the cost in time, energy, frustration and trees, not to mention the mental drain of doing and re-doing wasted, useless work, was huge.

Unfortunately, you can rarely get away with saying, "No, I don't believe you," to your bosses or project leaders whenever they ask you to trust them.

How can you prevent—or minimize—this type of problem? Two tools: reputation and knowledge.

Reputation and Knowledge

The better your—or the writing group's—reputation for meeting deadlines, the more control you will have over your own project destiny. If you are known for being the one who delays ship dates because the manual isn't ready, then you will be railroaded into finishing the docs before all the information is really there. Your work will be less than wonderful, and you'll get blamed for that. On the other hand, you might just be fired. The lesson here is: don't delay shipping dates.

Fortunately, in software, this is actually easier—but far more frustrating—than it sounds. In ten years of software development, I have never personally seen a product ship date delayed by the documentation.

Knowledge, best gained by experience, is what lets you calmly deal with the pressures of an approaching ship date. You will be able to direct others' attention away from forcing you to write based on hearsay and rewrite later, and toward the true Critical Path[40] for shipping. The more you (and your editors or boss or whoever) know about the development process, the better. You can be a resource that helps the company meet the ship date or you can be the scapegoat when the product slips.

On almost every product I've worked on, I've been to a review or scheduling meeting where somebody says that they're worried about the manual being finished on time, or that the manual is the Critical Path. The first few times I heard this, I worried about it. But all I could do was hurry up and wait for the needed information so I could finish. Or I could go ahead and finish on hearsay information, and end up redoing it. After a while, my response to people worrying about the manual was, "If the manual isn't finished, it's because the program isn't finished. You can't write about something that isn't there. If we really want to ship on time, I'd suggest that we take a look at the program to see why it is still not ready for feature freeze. And I'll bet anyone in this room lunch that the manual will be done and printed by the time the program is ready to ship."

If you know what you're doing, understand the schedule, do your job, work hard and fairly quickly, and above all, ask for help in time when you need it, you can get the documentation done on time. Unless you're lazy, incredibly slow, understaffed or just inexperienced and without someone there to help you out, you will do OK.

If the documentation isn't done, it's usually because the product isn't done. Once the product is feature-complete, your manual can be completed. After that, there's always some final testing and bug fixing on the product that will give you the time to edit, lay out and print the docs and deliver on time.

---

[40] *Critical Path is a project-management term for the series of tasks in a project that MUST be completed on schedule to ship the product on schedule. If any task that is considered to be in the critical path is delayed, a lot of people get worried.*

Layout and Graphic Arts

"Art for art's sake. Money for God's sake."

— Eric Stewart and Graham Gouldman

This chapter covers preparing a manuscript for layout, the layout process, dealing with graphic artists, and related topics.

The Basics

The Writer and Graphic Arts

Unlike in fiction writing, magazine writing or screenwriting, technical and UnTechnical writers have to think about how the words (and pictures) will look on the page. That means dealing with a graphic artist or being a graphic artist.

There are situations where the writer has no contact with and little influence over the graphic artist. You just hand off your text and graphics files and leave everything up to others. You can still help shape the final document if you organize and document your work well.

Your writing is the interface to the product. The graphic layout and binding of the document is the interface to your writing. The fonts, the font size, the spacing, the white space on the page, the placement of the graphics, the location of the index and the whole design can influence whether your work will be read and put to good use, or be stuck on the bookshelf, read only by dust mites.

When writing and preparing a document for layout, think a little about your readers:

- People read in different ways: straight through, scanning, just headings, or just pictures and captions.

- Organize the document with lots of headings and subheadings so readers can zero in on the information they need.

- On complex graphics that impart a lot of information, ask the graphic artist to make the graphic large enough for the detail to be seen.

- Make it easy on the reader by requesting a distinctive type style for instructions they will follow, as opposed to explanations or commentary.

Document Formats

Unlike in the world of screenwriting, there is no set universal format for a technical document. There are standard formats within each company, but those standards are (hopefully) designed and continually updated to meet the needs of the company's and product line's customers.

If your company has standard formats, get familiar with them and write to them as well as you can while still providing a good document for the customer. If the format is well-established and you think it needs changing, suggest small changes here and there to test the waters— and to find out if someone has considered similar thoughts and rejected them for a good reason.

When designing a standard format for your company, think about the reader's needs, as well as your own, graphic arts' and production's.

Above all, remember that a technical document is not only a linear presentation. Most readers will access bits and pieces, out of order, as it suits their varied needs.

The Writer as Graphic Artist

I can't go into the actual details of design and layout in this book, but you may at some time be expected to lay out your own work. So learn a little about design basics, and know your program(s).

The more popular word processors today are capable of a lot, but a dedicated page-layout program has many advantages when it comes to setting up grids, flexibility and dealing with color.

You need to stay compatible with your clients. Find out what software they use, what their vendors prefer, and learn it.

If you're really familiar with a genre of program, you can pick up a book on another program in that genre and get the general gist of it without buying the software. See if you can sit at a friend's house or a client's desk for a little hands-on experience. Don't buy an expensive program unless you know you'll need it and will earn your money back.

It also pays to have friends who are graphic artists who can advise you or trade their time for your time and help you out now and then.

Sometimes you may be given the PageMaker or Quark or FrameMaker file to input edits and changes. Check the page breaks before and after the area where you typed. Adding one word on a page can re-flow the text for the next dozen or more pages. It only takes a few seconds to fix and re-flow, but if you don't check it, the book could go out in horrible shape.

Dealing with Your Graphic Artist

First of all, the people who do layout are not "layout people." They are graphic artists. They like to be referred to as graphic artists; their department is graphic arts, not layout. Layout is a process they perform in their graphically artistic way.

What's in a name? Well, as a writer, do you like to be referred to as a typist? It's a matter of respect and politics. Show the respect, and you'll get more cooperation and better results.

Second of all, graphic artists are not typists. I mean this both in the sense that typing is not really their job, and that any time I've known a graphic artist to make changes that involved typing more than a few letters, there's been a typo. Personally recheck any pages where that artist has had to type in more than two letters.

A good graphic artist has a background in art and design, plus a thorough knowledge of various layout and graphics programs. Bow to their expertise when you come to an impasse regarding design or graphics use. If nothing else, that will help them bow to yours when it comes to the words. When working with a graphic artist, try to stay flexible. Give them the benefit of the doubt—after all, their main concern is making your work look better so it'll be easier and more enjoyable to read.

The graphic artist is responsible for the way everything looks on the page—for making it pretty. And that's very important. It's easier to read something that looks good than something that's either jumbled, or is solid text. But since the graphic artist is solely concerned with the look of the book, and is always on a tight schedule, don't count on them reading the document. They work with what's there. You will have to look through it to make sure all your notes have been removed.

You will have to make sure that the right graphics, headings and captions are in the right place.

It happens less often in a large technical document than in shorter pieces or marketing pieces, but sometimes a graphic artist may ask you to shorten or lengthen a sentence. It's amazing how adding or removing one word will make a whole page or chapter flow better and look better. Give it a try. Don't get high and mighty and demand that your prose be untouched. Then again, don't sacrifice the quality or meaning of the text. If you can't say it correctly and cleanly in more or less space, then don't. Tell the graphic artist you tried, but it just doesn't work. They'll understand and find a compromise. Trying is what will keep the graphic artist on your side. But succeeding is the sign of a good, flexible writer. Most languages, English included, allow many ways to say things.

Process

Preparing a Draft for Layout

You're done with your writing. It's been edited and polished. Now what? Time for layout. So you email the text and graphics to the assigned graphic artist and head off for a well-earned vacation, expecting to see a perfectly finished product when you get back.

Not likely.

The way to have everything go smoothly through the layout process is to properly and carefully prepare the manuscript for layout. You'll ask questions and write various notes, lists and explanations. Don't worry, it doesn't take long. And most of this prep work can be done as you write if you keep at it.

Here's what you turn over the graphic artist and how to do it:

- **The draft itself, electronically, in the proper format**

Check to see what format they want. It may be the standard output of your word processor, an RTF file[41] or a plain text file. The newer layout programs will import the standard word processor files cleanly, without losing information, but check to make sure what the graphic artist really needs.

This draft should include no graphics—just text and calls for the graphics. And beyond the file format, check in advance how the graphic artist would like the headings and styles for captions, instructions, notes, etc., to be assigned. If the import works properly, then you can go ahead and assign styles to everything in your word processor. If the import loses information, then mark headings and styles right in the text: [H1] for heading level one, [H2] for heading two, [instr.] for instructions, etc. If your word processor and the graphic artist's page-

---

[41] *RTF stands for Rich Text Format, a nearly universal text document format that almost every word processing and page-layout program on every platform can read and write. It can be indispensable for transferring documents between programs and computer types. It is a major step up from simple text, containing font and paragraph formatting information, but doesn't contain all the formatting, graphics or page-layout information that today's word processors and page-layout programs' data files contain.*

layout program work well together (and most do today), then assigning the headings and styles will save graphic artists a lot of time. They can simply redefine the look of each of the styles and start laying out without going through the text line by line and making sure each paragraph has been assigned the correct style.

One more thing about file format: many graphic artists work on Macs, while many writers work on PCs. Make sure that you have a seamless way of transferring files via disk or network that won't cause compatibility problems. Unless it's a brand new company with novices in all positions, somebody there will be able to tell you what works.

- **A printed copy of the draft**

This is the reference that the artist (and you) can use to make sure the headings and styles don't get mixed up. It also saves a lot of time. Often the graphic artist will delete a call for a graphic, then import the graphic. Sometimes they get distracted or lose their place, or the computer crashes. Having the hard copy nearby is a reminder of where they are and where those calls for graphics (or other notes) were.

- **The graphics in electronic form**

Before you start creating the graphics or taking screenshots, find out what format(s) the graphic artist prefers and what other formats are acceptable. Depending on the tools available you may or may not be able to supply the graphics in the preferred format. Chances are your graphic artist has a way to batch-convert whatever you hand over to their format of choice, but if you can give them what they need, life will be better for both of you.

If there aren't many, the graphics can be supplied on any number of disks—from floppy to multi-gigabyte—that are interchangeable between computers and even between Macs and PCs. If the graphics are too large or too numerous, or if the in-house process prefers it, then transfer the files via the network. Most graphic arts departments will have a place on the network for just this purpose. There might even be a specific "in box" for each project or for each graphic artist.

- **Printed set of graphics**

It may prove helpful to supply a printed set of the graphics as well. Print many to a page, along with their file names. Especially if you give your graphics files numbers for names, graphic artists will have no clue whether they're placing the right graphic. Having a reference sheet that they can check every so often eliminates potential confusion.

You should also mark up the printouts to show how captions and callouts should be shown, as well as any other special features or graphic manipulations that you want.

- **Graphics list**

The graphics list, originally created for your use while creating the graphics, may contain notes for the graphic artist about special handling of certain graphics. And if you added a "GA Use" column, they can use it to track their progress and completeness.

A sample graphics list can be found in the UnTechnical Writing chapter and in the Exhibits section.

- **Style sheet**

This is just a brief list of the different headings, subheadings and type styles that you've defined in the document, along with a brief explanation of the purpose they're serving. You can suggest comparative sizes of headings, and request that instructions stand out clearly from the rest of the text, but leave actual point size as well as fonts up to the artist.

Here's a sample style sheet for this book.

- **Style Description for UnTechnical Writing**

Heading 1 is for chapter names only. It only appears on chapter title pages, which preferably only appear on odd-numbered pages. Big, beautiful, arty.

Heading 2 is a major section within a chapter. It would be great if it could always push to a new page. There is no body text between the Heading 2 and Heading 3, so some space or a graphic element may be needed. Smaller than Heading 1, but arty is OK.

Heading 3 is the most common—and possibly most important—heading level, and needs to stand out very clearly when the reader is quickly paging through the book. Try to keep it large, with wide kerning. This heading will appear in the contents. No need to push it to a new page.

Heading 4 is a minor subhead, and should stand out from the body text, but needn't be very large, since it won't be in the contents and won't be something that readers would search for.

Body (or Normal) is the majority of the text in the book. Easy readability is important. Not too small, please.

Bullet is a bulleted version of body. It needs different indents at different times.

Instructions will appear occasionally within body text. It should stand out from the body text, but not totally interrupt the flow. Bold, possibly sans serif.

Quotes and Quoter are for quotes and the person credited with it. They appear on chapter title pages, and are sprinkled throughout the book.

Footnotes should appear on the page where they are called, not at the end of the chapter or end of the book. Please make them readable without a magnifying glass.

Headers and Footers—your call.

Other—I have a few slightly modified lines here and there. For instance, a short line may be bolded and centered or the first lines of a series of paragraphs might be bolded. I haven't marked these as different styles. Some of this could be made into a Heading 5, but there aren't many, so it probably isn't worth the effort.

- **Cover letter/notes**

This is hopefully only a page or so containing your contact information in case there are any questions, plus any other comments and suggestions. It's best to go over this in person if you can, but hand it over in writing as well.

Oh, and one more thing: **If it's a software project, keep a set of the graphics in electronic form.**

After you've delivered all the graphics to graphic arts, you'll be tempted to wipe them out, freeing a lot of space on your hard drive. But if it's a software project, and you (or someone else) will be creating an online help file, you'll need the graphics in their original, full-color form. If you forget, you may be able to get them back from graphic arts, but sometimes the first thing the graphic artist will do is batch process all the files to change the format and convert them to grayscale for printing in a non-color manual. You'll want the color shots for the online help, and once color's gone, you can't get it back. At that point, the only thing to do is to use grayscale graphics in the online help (which won't match the program), or retake all the shots you need.

The Pre-Edit Edit(s)

Long before the layout goes to an editor, and at the graphic artist's convenience, give it a once- or twice-over.

As a minimum, check that:

- The import process didn't lose anything, including tables, bulleted lists and numbered lists.

- All notes, questions, calls for graphics, brackets, etc., have been removed from the document.

- All headings are at the right level and in right place.

- All lines and paragraphs have been assigned their proper style: heading levels, quote, instruction, etc.

- All the artist's styles make sense for their purpose. (Don't redesign the styles, but make sure the artist knows what function the text serves. It makes more sense for instructions that the customer follows be large, clear, simple and bold than thin and fancy script.)

- All the graphics are in the right places and are as clear as possible for a test print.

- All the captions and callouts are in the right places and the lines from the callout text to the graphic point to the correct spot.

- Page numbers are continuous, and that any headers and footers give the correct title and chapter names.

- The text flows properly from page to page, with no paragraphs or parts of paragraphs accidentally hidden.

- There is a consistent look from page to page and chapter to chapter.

And while you're checking things over, you might as well mark up the draft to indicate what you want in the index.

Indexing

I put indexing into the layout section because it is usually done in the middle of the layout process. If you and your graphic artist use the same program for writing and layout, then you can create the index earlier. If that's the case, then pretend this blurb on indexing is in the writing section.

Before you can create a useful index, the text and graphics must be in the final page-layout program. Indexing rarely imports between programs.

If the document is a large, complex book, then you may want to get a professional indexer to do it. If it's very small, then you may be able to get away with a simple, single-level index. If it's in between, then you can usually start with a single-level index, and regroup the index points and give them group headings to make a still fairly simple two-level index. Different programs will have different built-in indexing tools, so check with the local expert on the program to find out the best local process.

The easiest way to do an index is for the writer to mark up the first rough layout, indicating the words and phrases you want in the index. A highlighter works well for this, and distinguishes the index selection from those red pen corrections. From this draft, the graphic artist will create a first draft of the index, with the tools built into the layout program. The writer then edits this draft, changing the names of things when it's helpful, grouping similar subjects under new headings, and making capitalization, fonts and font properties consistent.

Always recheck the index, both the content and the page numbers, during final edits.

The Rush Job

Sometime in your career (if not every time) you will hit a deadline so hard it throws the whole company out of alignment. Suddenly, reasonable people are running in circles in panic, and demanding the impossible.

Other than getting your writing done on time, the only other way you can help speed up the process and push for the deadline is to work even closer with your graphic artist and editor.

The first thing you can do is provide FPOs (for position only) for graphics that aren't final. The graphic artist can use these to get the layout as finished and polished as possible, then drop in the real graphics when they're ready. Be careful to supply FPOs in the proper size or it'll require a lot of pages of rework. And be sure all the FPOs are marked, listed, and eventually removed.

If the document isn't finished, divide it into parts and forward those parts that are finished and edited, so the graphic artist can work on them separately. You may be tempted or asked to send the parts that aren't finished and edited to layout as well, but that's always a mistake that requires a lot more time and energy in the long run. Making edits and major changes during the layout process takes much longer and involves more people and ridiculous logistics.

Be ready to sit with the graphic artist (sometimes all night) and set up a bucket brigade of pages. The artist will input the changes from the editor, and print out each page—and the following page—as it is done, and then go on to the next page to enter edits. You check the newly printed page against the editor's edits to make sure they're right, and that they haven't caused other problems, like messing up the flow of the following page or losing part of a paragraph in layout limbo. You allow the graphic artist to work solely on correcting the document by taking over the duty of collecting all the final, finished pages and collating them into a completed draft that the editor can sign off on.

The ability to handle rush jobs and finish on schedule without panicking is the measure of the writer, editor and graphic artist. The ability to avoid rush jobs is the measure of the producer, project manager and the company atmosphere.

UnTechnical Layouts

Ideal layouts for UnTechnical documents are very similar to good layouts for any documents; they should be well-designed, good looking and above all, be designed for the reader, not for the graphic-arts contest judges.

Since we don't want to scare or intimidate the reader, the layout shouldn't look too serious (like a typical computer manual written for programmers). We don't want it to appear too lightweight, either, so the overall look and feel should be somewhere between serious and frivolous.

Leave lots of white space on the page, with reasonable line spacing and extra spacing between paragraphs to emphasize the bite-sized chunks of writing.

There should be lots of graphics—useful graphics, that is. Don't just throw in graphics for the sake of graphics unless they help the reader understand the subject matter or really help the look of the document. Use a series of graphics showing steps or processes when it's helpful. Graphics should have useful and easily readable captions and callouts.

Headings and subheadings should be extra large—almost to the point of parody—so the reader will read them even without trying while casually flipping through the book. Later, when they need to know something, they'll often remember that related heading and look for it in the manual instead of calling the tech support line.

Definitions for difficult, technical or product-specific words should be placed on the page where they are first used, either in the margins or sidebars, unless they're written right into the text. Collect and duplicate the definitions in a glossary at the back of the document as well.

Put the index as close to the back page as possible, so it can be found easily or even by accident.

Choose font sizes and the fonts themselves based on the readers. People of different ages require different type sizes (larger for oldest and youngest). Younger readers may appreciate fun fonts. But make sure they're easy to read and mostly use them for headings, not large bodies of text.

Interface Design

"The higher we soar on the wings
of science, the worse our feet
seem to get entangled in the
wires."

— Unknown

This chapter introduces you to the basic concepts involved in designing
interfaces for products—hardware and software—so they are better-
suited to humans, even and especially nontechnical humans. This
information can be used to contribute to design, but is here mainly to
help you better understand what you are writing about.

The Basics

Introduction to Interface Design

If the world were a perfect place, interface design wouldn't be a separate step in the design process. It would be integrated into the whole product-design process. In that spirit, some of the following guidelines will apply to designing the whole product, not just the interface.

In reality, most technology-based products, especially software programs, are first engineered to function and then an interface is slapped on. And the writer is usually brought into the project in the middle of the slapping to explain the not-yet-complete-much-less-tested interface to the customer.

Big projects and big companies may have a team of experts to do the interface design. It is best done with a team of industrial designers, cognitive scientists, psychologists, engineers and others … but in smaller companies, it's usually just done by whoever is around. If you're around, then you should know something about it.

The more you, the communicator, know about interface design, the better you can explain interfaces to your readers. And if there are experts around, then you'll have the vocabulary and basic understanding to communicate with them. You might even help to shape the product to better suit the customer.

Interface design is a big, big subject. This little section of this little book will by no means make you an interface designer. Nor will this book describe the fringes of interface design, which include everything from virtual reality to voice control to gesture interpretation. But it will give you the basics that every writer working on technical material should know.

You may notice that some of the information in this chapter will sound familiar—almost as if you've read similar things recently. You probably have—earlier in this book. Writing for the nontechnical reader has a lot in common with designing a product interface for a nontechnical user.

A last introductory note: interface design is a process, not a fixed set of rules. There are rules of thumb and guidelines that you can use, but no set path to follow. Each project is a separate journey. Every solution is a compromise between the product's technical capabilities and the customer's human capabilities. Nothing will ever be perfect, but you can eliminate a lot of the problems.

> "Technology is ruled by two types of people: those who understand what they do not manage, and those who manage what they do not understand."
>
> **— Attributed to Mike Trout, also known as Putt's Law**

Writer as Interface

If you write product descriptions, manuals or instructions, you are the interface between the customer and the concepts, ideas and information necessary to use the product. The way you present the words, the words you choose, the order you put them in, and how well you use or abuse the conventions of language and grammar, is all part of that interface—the part that you design.

The graphic presentation of your information, whether on paper or onscreen, is another layer of interface that needs to be designed, sometimes by you, sometimes by others.

In order to be a better interface, you have to think about interface and design. Look around you. Everything anyone makes has been designed. Often badly. Learn from the bad as well as from the good. To get started, read any and all books by Donald Norman.

As a writer, there are a couple of things to watch out for:

- **Hard-to-explain interfaces**

A major test of an interface design is how easy it is to explain to the customer. If it's easy to explain, it's easy to understand.

If it takes you three pages and four diagrams to explain what a particular button does, then maybe there's something wrong with the way the button or the function behind it was designed.

If you're involved in design decisions or around when they're being made, try to make sure that the solution that is easiest to explain is seriously considered. It's usually best for the customer, and it's better for you.

- **A big glossary may be a sign of interface problems**

If the product is very, very technical, or a truly new type of product, a large glossary may be necessary. But for a not-so-technical, familiar type of product, if you need a huge glossary, take a close look at the interface. It could be in trouble.

Of course, even in a fairly simple, well-understood product like a television, a glossary (defining brightness, contrast, comb filter, lines of resolution, etc.) may be good for added value. But the customer shouldn't need to constantly refer to the glossary to understand the product.

> "All technology should be assumed guilty until proven innocent."
>
> — **David Brower**

Thinking Like a Designer

Here are a few points to absorb that will help to give you the right mindset for interface design:

You Know Too Much

You are not the typical customer. You know too much. If you are technical enough to design or write about a product, chances are you are far more technical than your target audience.

Even if the customer is very technical, the fact that you know the product intimately and the customer is approaching it for the first time sets you apart.

You need to test your product and the interface with real typical customers, and not just guess, assume or dry-lab.

Know the Customer

Understanding the customer goes beyond interface design to the very heart of product design and selection. When you design something, you're designing it for someone. If you don't have a good picture of your customer, you can easily do a bad job.

Does your customer have special needs? Deaf? Blind? Three years old with tiny fingers? These physical aspects of the customer are important for you to know. As mentioned earlier in this book, you should know as much as you can about your customer, including:

- Age and age distribution
- Gender and gender distribution
- Technical expertise level
- Education level
- Percentage of repeat customers
- Professional fields
- Other attributes, including cultural.

Be the Customer's Advocate

You, as the writer, may find yourself on a small design team responsible for an interface, along with a programmer/engineer and a graphic artist. Stereotypically, the programmer or engineer is concerned with the thing working and with getting

Example of a Cultural Mismatch

A great example of a cultural mismatch between a product and the intended customer was the Julie Doll. A toy company made a fortune selling a high-tech teddy bear with a tape recorder in its belly. When you played a tape, the bear would move its mouth as if it were talking. It could simulate reading a story to the child. This was the hot toy one holiday season.

So far so good. Now it was time to follow up that success. Hmmm ... high tech made millions, it follows that higher tech will make even more. (Yeah, right.)

Their next big project was the Julie Doll. Never heard of it? It was a technological wonder. It was a doll with speech capabilities and all sorts of sensors. If you picked her up, she'd say, "Where are we going?" If you put her in the refrigerator, she'd sneeze. If you covered her eyes, she'd comment on how dark it was. And she had speech recognition. When you turned her on for the first time she would guide the child (or whoever was around) to repeat a number of special words, and instruct the child to use those words in a conversation. It was great, or eerie, depending on who you were. If you slipped those words into questions or comments, it really seemed as if the doll understood you and carried on a conversation.

continued

as many features into it as he can. Stereotypically, the artist is concerned with making it pretty. When faced with these two stereotypes, the writer must be the advocate for the customer, and walk the line between artist and programmer, tempering them both with the user's needs.

Let the Customer Interact With the Task, Not With the Product

People buy hammers to drive nails, not to play with the hammer. People buy VCRs to record and watch movies, not to mess with a VCR. People buy application software to perform a business task, not to spend hours fiddling with their computer trying to figure out how to make it do what they want.

Think about what people want to do, and how they would most likely go about doing it, then design the interface to allow them to work their way.

VisiCalc was the first true "killer application," a computer program that was so useful that it was worth buying a computer just to use it. It wasn't perfect in function or interface, and compared to today's spreadsheets, it's practically Stone Age. But it allowed people to do a task they needed to do in a way that they already understood. It was a familiar ledger sheet—except that it did all the calculations for you, and did them quickly and accurately.

Make the Interface—and the Product—Disappear

Taking the previous point one step further, if your product and interface are designed really well, they will seem to disappear. The customer will be thinking only about the task, and not about the product at all. After as short a learning time as possible, the user stops noticing the tool and just uses it.

Think of using a hammer. After the first time or two in your life that you see or use a hammer, you don't really think about the hammer. You think about that nail you're driving into the wood.

Or a typewriter. Once you learn how to touch-type, when you sit down to write or copy something, you think more about the thing you are writing than you do about the tool you are using.

Function Over Art

Design for functionality, not art. It has to look good, of course, and clean, and inviting and comforting, but it isn't a palette for free artistic expression. There's an art to it, but it itself isn't high art.

Here's an example. A friend of mine received a beautiful tea kettle as a gift. It was visually stunning, and the box mentioned that it is on permanent display at some fancy-shmancy museum. But the tea kettle is unusable as a tea kettle. Not only does the handle get hot enough to burn you, but it forces your hand into a position so that whenever you pour boiling water out of the kettle, rising steam scalds your hand. I'd like to see the artist who designed it use it (and I'll hide the burn ointment).

> "The only good thing I can think of about computer interface designs is that the Museum of Modern Art doesn't yet give prizes for them or exhibit elegant examples. Like they do for watches and toasters and fountain pens. Once they start to do that, that is the end of functionality."
>
> **— Donald Norman, from *The Art of Human-Computer Interface Design*, edited by Brenda Laurel**

I had the chance to witness some accidental focus groups on the Julie Doll. When the toy company went out of business (in part, but not completely due to losses from the Julie Doll), someone from that company came to work for the software company I was working at. He mentioned the Julie Doll, and a number of us techies were interested, so he went back and picked up a few factory seconds scheduled for destruction and brought them to work. The results were amazing.

Almost universally, the results were that men loved the doll, and had a great time testing all the sensors and trying to evoke all the possible reactions. Also almost universally, women hated—or feared—the doll. I saw grown, intelligent women hold their hands in front of them, making a cross of their two index fingers, and run from the room as if a demon from hell had appeared in the form of a doll. (Of course all those Chuckie movies didn't help things.)

Young girls, the intended audience, didn't really care about all the high tech stuff. They had no intention of putting their doll in a freezer, and spent more time combing the hair and playing with the doll as if it had no built-in techno-goodies.

So the mistakes that the company made were: the intended audience didn't like

continued

Real World Considerations

In the real world, you rarely have the time, team and budget to design the product or interface as well as you can, should, or would like to. You have to work to a schedule and within a budget. Time and cost usually win out over usability. You just have to do the best that you can.

Even if you do a wonderful job, eventually the demands of the market will ruin it. Suppose a company makes the perfect widget. All the other companies in the world that make widgets won't stop making them and go out of business or make something else. They'll make new, improved widgets with extra features that will indeed be extra, but won't necessarily make the widget any better. An example of this is the very word processor I'm using to write this. The latest version has 20 or 30 new features, maybe one or two of which I might actually use. Of course, five or six of the older features that I liked no longer work, work as well or work the same way. Chalk one up for improving a good thing until it stinks.

Often the client isn't the user. A company may order a design with complete specs and requirements. And those specs and requirements may not have the end user in mind.

The world of business doesn't reward good interface design. Many companies buy entirely on a cost basis, totally ignoring usability.

Often, designers are isolated from the customer by the very people that are charged with learning what the customer wants and needs: the marketing department.

The Balancing Act

All these things—and more—go into design:

- Aesthetics
- Usability
- Cost
- Ease of manufacture
- Safety

Good design is a balancing act, applying all of these criteria to every decision. Problems arise when one aspect dominates all the others in a design.

Keep your balance.

> "Technology is a way of organizing the universe so that man doesn't have to experience it."
>
> **— Max Frisch**

technology for technology's sake, and would have been just as happy with a low-tech doll that would have cost far less to develop, and would cost less in a store.

While the child is the actual end user of the doll, it is usually the mother who buys it. And most of the women in our informal test group wouldn't allow that thing in their house. And there was no way they were going to buy a doll that costs over $100 for a four-year-old.

The only potentially successful audience for the doll was men, but most men don't buy themselves dolls.

Now, only a few years later, there are a number of very high-tech doll-like toys on the market. And they seem to be doing quite well. What's changed?

Even though it's only been a few years, technology has continued to work its way into our homes and offices, and has become more accepted by parents.

Because the technology has grown, it is both better and cheaper than five years ago.

These newer toys are better designed, and are designed for the actual audiences (for the kids and their parents, not for technogeeks).

Most of these toys don't appear to be human, taking away the eeriness and the "Chuckie" effect.

Times change, cultures change, technology changes, designs improve.

More Basics

Concepts

Here are some basic concepts that interface designers know and use.

Knowledge in the Head and Knowledge in the World

In order to use something, you have to know how to use it. Where do you get the knowledge?

If you can look at the thing and see how it works, then the knowledge is said to be in the world, or in the thing itself. If you look at the thing and can't see how it works, and you have to read a manual or instruction sheet and memorize a number of arbitrary facts, then the knowledge is said to be in your head.

For an example, let's compare doing a complex task for the first time on an old and a new word processor.[42] Let's say you want to insert a file from a disk into the currently open file.

On an old (circa 1982) word processor, you'd have to look up and type a sequence of control keys, then type in the name and location of the file you wanted to insert. Chances are the next time you needed to do it, you'd have to look it up again. On a newer word processor, you click on the Insert menu, then find File... in that menu, then point to the file you want to insert. Everything you need to know is there, or appears as you need it.

Yes, if you memorized the control key sequence for the help screen, the older program would remind you of the arbitrary control key sequence you needed. But you still have to leave the program to look it up each time until you memorize it. When the information is in the product, as in the newer word processor, you don't have to look something up, then do it—just finding what you want to do performs the task.

Whenever possible, put the knowledge that the user needs into the product itself, so a reasonable person can figure out how to make it

---

[42] *I'm comparing interfaces and information placement here, not judging products from different times and technology levels. In the time of the earlier word-processing program, computers had 64K or less RAM, and no hard drives. It may have been impossible to make a better program given the times and technology available.*

perform the desired task. It is good if the person can use the product more efficiently once they have some experience with it (and therefore knowledge in their heads), but let them at least be able to use it somewhat just by looking at, and possibly experimenting with, the thing itself.

The way to do this is through understanding and making use of the concepts of visibility, mapping, affordances, constraints and feedback, and being aware of the mental model you are presenting. All of these concepts are introduced below.

Visibility

Good visibility means that someone can look at the product or its controls, and tell what state it is in and what actions are possible.

For an example, let's look at a typical toaster. A toaster has four states: empty, ready to toast, toasting and done. You can tell what state it's in just by looking at it. If the handle's up and there's nothing sticking out of the toaster, it's empty. If the handle's up and there's bread sticking out of the top, it's ready to toast. If the handle's down and there's a red glow coming out of the top, then it's toasting. If the handle's up and you can see brown toast sticking out of the top, then it's done.

What about the visibility of the possible actions? The toast slots are fairly obvious places to put the toast. And there are two controls. One goes up and down. The other is either a rotating knob or a slider.

A toaster has good visibility. You know by looking at it what it's doing and what you can do to it. Examples of bad visibility are the old word processor mentioned above, and most modern office phone systems.

Mapping

Mapping is designing controls so it is easy, or at least possible, for the customer to determine the relationships between actions and results, between the controls and their effects.

Once again, think of the typical toaster. Generally, the mapping on toasters is very good. The control to lower and raise the toast is well-mapped to the movement of the toast itself: push it down to lower the toast, pull it up to raise it. (There is also very good feedback on this control, since it is usually physically connected to the platform holding the toast.)

The control for toast darkness is either a left-to-right slider or a rotating knob that maps very well to a continuum of time and toast darkness: left or counterclockwise for "short time" or "light," and right or clockwise for "long time" or "dark." (Of course, in countries where people read from right to left, this may be a little less intuitive.)

An example of bad mapping is a typical digital watch. There are three or four buttons which must be used in combination and sequence to set the time, set, activate, deactivate and turn off the alarm, change modes to stopwatch and other actions. Is it obvious from looking at the buttons which one does what? Can you pick up any digital watch and run it through its paces without reading the manual? After you read the manual and don't use the watch for a month, can you still work all the functions without looking them up again?

Affordances

Affordances are the properties of things that tell us, visually or otherwise, what they do and how they work.

The classic example of this is door hardware. A door knob, by the way it looks and the way it feels, tells us that we should turn it. A flat plate on a door, by the way it looks and feels, tells us that it should be pushed. Other types of door handles, such as those on car doors, tell us to grab and pull.

Problems arise when designers ignore affordances. How many times have you tried to pull a door that should be pushed or pushed a door that should be pulled? Don't feel bad. It's not your fault. The person who chose the hardware doesn't understand affordances. Sometimes, these designers notice that people have a hard time using their doors. So what do they do—change the hardware? Of course not. They put the word "push" or "pull" on the door—a user manual for a door. It's only one word, but it's unnecessary if you choose the right hardware.

The doorway to the boardroom at a software company where I worked is a prime example of ignoring affordances. The door handle has an affordance for pulling, but the door must be pushed to be opened. People pull, and nothing happens. So they pull harder. The handles are always loose from the wasted effort of being pulled so hard and so often. These same handles are on the inside of the door. Nobody ever has problems opening the door from the inside; the handle tells them to pull, and they do.

Constraints

Constraints are ways to limit and lessen the number of possible actions, to make something easier to understand and use.

A good example of this is the scissors. Try cutting a complex pattern in paper with two unattached knife blades. It should work—after all, what is a scissors but two knife blades? But a scissors is more: it's two constrained blades, that can only be moved in limited ways designed to cut paper. It's also constrained in how you hold it. If you grab the wrong end, it lets you know. And it won't work.

VCRs, for all their problems, have eliminated one by constraining the tape cartridge so it can only be inserted into the machine correctly. You can't put it in upside down or backwards.

Feedback

Feedback in design is returning to the user information about what they just did. Without feedback, it is easy to get lost. Try to draw a picture on your computer using your mouse and a paint program—without looking at the screen. That's working with no feedback.

The best feedback is visual. You should be able to look at a button or switch and see how it is set. This goes for virtual buttons on a computer screen as well as physical ones.[43]

Audio feedback is also important:[44] when a switch is flicked, you hear it change state. Older computer keyboards had a nice click when you hit the key. Even when you typed fast, you'd know by the sound (or lack of it) if you missed a letter. Telephones give audio feedback; you can hear whether or not you've pressed a button and you can tell the state of the phone call by sound (dial tone, ringing, busy signal).

There is also tactile feedback. My old keyboard has wonderful tactile feedback. I've used the same one for 10 years, switching it from com-

---

[43] In the U.S., the standard for a light switch is up for on. In Europe (at least the parts I've been to), down is on.

[44] Be careful with sound. As powerful as it can be for feedback, it is easy for it to become too cute, too annoying.

puter to computer. I can tell if I miss a letter by feel. Besides that, the keys have a good, solid (but not hard-to-push) feel. All the new keyboards I try feel like sloppy mush. If you can get a hold of an old Northgate or old IBM keyboard, try it.

Always try to give the customer complete and continuous feedback on the results of their actions.

Mental Models

People understand how things work by creating mental models. The models we make are based on preexisting knowledge—whether or not the knowledge is accurate.

There are actually three mental models involved with any product:

- The engineer's model—the true, technical inner workings of the product as understood by the person who engineered or programmed it.

- The interface model—the mental model that the interface designer is presenting to the customer to explain how to understand and operate the product.

- The customer's model—the mental model that the customer forms, based on what they learn (and don't learn) about the interface model and what they already know (or think they know) about the world.

All three models can be very different. Ideally, your interface and the customer's models will be similar, but when they're not, you've got problems.

You need to understand how the customer normally goes about a task (and what they know or believe about it) before designing a new interface for that task. If you build your interface model to work with the customer's existing models and knowledge, you'll have a happier customer and a better-used product.

In other words, make it work the way people expect it to work.

> "... what's the point of ... new technology if you can't find some way to pervert it?"
>
> — **G. A. Effinger**

All About Errors

A big part of interface design is detecting and eliminating errors that the customer might make.

There are two main types of errors: slips and mistakes. Slips occur when the user is trying to do the right thing, but doesn't. They've formed an appropriate goal, but didn't do the right thing to reach it.

Mistakes happen when the user forms and goes after the wrong goal. Mistakes are generally worse than slips, both in the quality of design that caused the mistakes, and in the results.

Slips

Slips are most common when people are bored, distracted, not paying attention or under stress—all common traits of people trying to learn about a new technical program or gadget.

Here are some types of slips, with examples.

- **Capture errors** are errors caused by falling into old habits; having your actions captured by common actions in a similar situation. For example, say you are using someone else's pen, and by habit, you put the pen into your pocket. It's not necessarily kleptomania. Your actions in this different situation were captured by your usual actions in the typical situation. You usually use your own pen and put it in your pocket when you're done, so, unless you consciously remember that it is someone else's pen, you will likely put their pen in your pocket as well. Whether the bankers know it or not, all those pens attached by wires or chains aren't preventing petty theft, they're preventing capture errors.

- **Description errors** are when different actions have similar descriptions, causing someone to do the right thing at the wrong place or to the wrong object. They often occur when the right and wrong objects are right next to each other. For instance, grabbing the vinegar bottle instead of the wine bottle and pushing the button on your phone for line 1 instead of line 2 are both description errors. Car fluids are made in different colors and their filling compartments are all different shapes and sizes to help prevent description errors. Modern audio and video equipment with many identical-looking and -feeling buttons that do different things all placed in a row are designed to cause description errors.

- **Data-driven errors** occur when you mix up your data. You're typing with the radio on, and find yourself typing what the radio announcer said. You're talking to someone while thinking of someone else and call your companion by the wrong name.

- **Associative activation errors** are what Freudians live for. They're when you do something inappropriate because of an association between what you want to do or say and some internal or external event or thought. You pick up a ringing phone and say, "Come in," or you make any of a million Freudian strips ... I mean slips.

- **Loss-of-activation errors** are when you forget what you were doing, or do something and forget why. What was I just saying?

- **Mode errors** happen when the user thinks the product is in one mode, but it's in another. The most classic and most drastic cases of mode errors involve airplane autopilots. For example, the pilot looks quickly at the display and sees the number 32. It could mean the autopilot is set to descend at a rate of 3.2 degrees or to descend immediately to 3200 feet. The only way to tell which mode the display is in is by a small indicator light elsewhere on the autopilot unit. Confident that it is in degree mode, the pilot and plane plow into the 4,000-foot-high mountain.

Mistakes

As stated above, mistakes result from having the wrong goal. Mistakes are often caused by designs that give wrong or unclear messages. Incomplete or misleading information can cause people to make poor decisions, misclassify a situation or fail to take all the relevant factors into account.

Mistakes are also caused by the fact that people rely on and act on memory instead of analyzing the current situation. They do what they did before in a similar situation, which may not be similar enough.

One of the most common mistakes designers make that causes customers to make mistakes is to assume that all customers will act analytically and rationally. No offense intended, but humans in general, in their thinking, in their problem solving and in their planning, base more decisions on past experience than in logically looking at the present.

To compound the problem, when you design something or evaluate a design, you are generally in an analytical mode. You are more likely to be logically looking at the thing and analyzing it. You and your customer aren't even on the same wavelength.

For this reason, designers and others involved in the process are very likely to miss big problems. They just don't see the product the same way as the customer. To find out how the end user will look at the product, you either have to kick yourself out of analytical mode for a while, or—the far better option—bring in typical customers for design testing.

Here's the part that's hardest to take: you can't blame the customer for making mistakes. Since long before the caveman days, we have relied on memory to save time and energy. If we had to think about how to walk every step we took, we'd never run marathons. If we had to think through every physical and mental action we do, we'd never get out of bed, much less get around to procreating the species.

The more that complexity enters our lives—and a lot of it is brought to our lives through technology—the more we have to automate parts of our lives to have the time and energy to use our brains for the really important things, like making clever conversation, making important decisions and writing books.

Technology writers, as spokespeople for the technology sellers, are adding more complexity to people's lives. Designers and design-team members need to accept that people are already overwhelmed by complexity, and will use memory over analysis whenever possible. We have to design and explain things in ways that make sense to people, in ways that fit in with their current mental models of the world.

Good luck.

> "Technology makes it possible for people to gain control over everything, except over technology."
>
> **— John Tudor**

Techniques

Here are a number of techniques and rules of thumb that will help you design better interfaces.

Keep It Simple

Don't make things more complicated than they need to be. Don't make the user learn new mental models if you can avoid it.

If you must have a lot of complexity (lots of buttons and controls), organize them well, grouping like functions together. It may even be a good idea to hide many of the controls, only showing them when they are useful.

Modes, Pros and Cons

Dividing the use of the product into modes is a way to simplify by only showing the controls needed in that mode. Of course, modes have their own problems, including mode errors.

Modes are a mixed blessing. They allow you to simplify separate functions and controls into smaller groups that work together, but they force the user into changing mental models for each mode.

Pay Close Attention to Names of Processes, Buttons, and Menus

Names of things are an important part of any interface. Don't let names be confusing or too technical. Don't name things by their internal electronic or software variable function; give them a name that will mean something to the customer. And don't let anyone get away with calling something a "Duration Interval Control Device" when something as simple as "Timer" will do.

Name things from the outside, from the customer's point of view, not from the designer's or engineer's point of view. This goes for all the buttons, screens, functions, switches and procedures that the customer sees, performs or uses. No code talk. No techie talk. Just human talk.

Learn From the Elders

Interface design, like writing, is a form of communication. High-tech hardware and software interface design is a relatively new form of communication. Look to the older forms of communication for lessons. In general, learning about writing, theater, animation, architecture, and even television can provide a lot of useful knowledge.

Test, Then Test, and After That, Test

User testing is not what you do after you're done with a project and want to make sure you're right or to identify a few little modifications to make it perfect. It starts at the beginning and lasts through the whole process. Test your product and interface with users at all stages.

Start user testing even before you begin your design, as a part of the task analysis.

Use Metaphors

By using metaphors in your designs and explanations, you can link your product and its features to the existing knowledge that your customer already has. Metaphors let you explain something new in terms of a customer's existing mental model.

The most common and basic metaphor in technology today is in explaining hard-disk organization in terms of file cabinets and file folders. The way a hard disk stores files is actually very different from a box of papers, but it can and does look the same to the user.

As with most things, it's easy to overuse metaphors and get too cute.

Memory and Standardization

Let's think about the toaster again. A wonderful example of good visibility and good mapping. It makes sense. It works. Why would anyone do it any other way? So toasters have existed for decades, and anyone who has ever used a toaster can use a new toaster just as easily. But leave it to modern technology to mess up a good thing.

About a year ago my wife and I needed a new toaster. We, as modern consumers, didn't just rush out and buy one. We first set the most important specifications: the number and width of slices. Then we researched what toasters met our specifications. From that list, we evaluated the other specifications to make our decision. We compared costs, colors, size, style, warranty, ease of cleaning. Then we bought one. So far so good.

We took it home, and it worked just fine. Until one day I got impatient waiting for the toast to pop. So I pulled up on the up-down control. Nothing happened. I tried again, with more force. After a minute of manhandling the poor toaster, I managed to force the handle up and get my toast. My analysis of the situation was that the toaster was defective. The up-motion of the control was somehow blocked or broken. It was only later (after enjoying my toasted breakfast) that I looked closely at the toaster.

On the up-down control was an icon with a picture of a piece of toast and two arrow heads pointing up. It looked like this was the proper way to raise the toast, but it also implied that this wasn't the way to lower it.

Below the up-down control and the darkness control was the barely visible word "stop." Looking closer revealed that the portion of the toaster below the word was depressible—a button that doesn't look like a button unless you look closely, labeled with white on white lettering that isn't visible from a distance. This was the up button. But it said stop, not up. And you wouldn't know it was there, or see that it was a button or see the word "stop" unless you looked very closely. And you wouldn't look closely until you found that the standardized way of working the toaster didn't work.

The standardized way that the toaster works is so ingrained in my memory, that it never occurred to me to check a new toaster, either when buying it or afterwards, to see if it worked differently. I resorted to checking the instructions, and it was clearly explained there. But I felt no need to read the instructions for using something that hadn't changed in function in 50 years, and that I had been using since I was five years old.

Sure, standards, like rules, are made to be broken, but there should be a good reason to break the standard. What if a large clock- and watch-maker suddenly decided to be different, and make all their clocks run counterclockwise? They'd sell a few as novelty items, but they'd lose market share. What if a carmaker decided to switch the positioning of the gas and brake pedals?

When you design, look at similar products. If there's a standard way that they work, or if people expect them to work in a certain way, you'd better have a very good reason to buck the trend.

Warning Signals

Warning signals can be a good way to warn people of mistakes and potential mistakes. Beeps, buzzes, lights and dialog boxes are all ways to let users know that they did or are about to do something wrong, dangerous or unproductive.

But warning signals can be overused or badly used, making them worse than useless.

The prime example of an overused signal is the alarm that tells you that you have your car key in the ignition with the door open. There are many times when you want the door open with the key in the ignition. But the signal is there and annoying. You get to hate the signal so much for bothering you when you aren't in need of a warning, that when you actually need the warning, you ignore it. And, of course, that's when you lock your keys in the car. It's the alarm that cried wolf.

While on the subject of car warning signals, I have a fairly new car with a computer-controlled sensor array. It checks an amazing amount of things in the car, and notifies the driver of any problems ... sort of.

One of the things that it checks is if the gas cap is properly in place. A good thing to check. But the signal the car gives when the gas cap isn't on right is to light the engine light. Not only does this tell you nothing about the gas cap, but it makes you think that there is something seriously wrong with the engine. Traditionally, when you see the engine light, it means trouble (standardization). The first time we saw the engine light go on, we called the dealer. They told us to check the gas cap, so we did. The light stayed on. There was no way for us to reset it to see if that was the real problem. We took the car in to the dealer, and they reset the alarm, and indeed it was the gas cap. The next time we saw the light, we checked the gas cap ourselves, and found it wasn't tight. But we had to take the car in to the dealer again to have the alarm reset, or we'd never know if there ever really was a problem with the engine.

A solution for this? Since there are dozens of sensors checking things all over the car, having a different warning light for each potential problem would be expensive and ugly and probably confusing. What would work would be to have a single red warning light, and next to it an LCD display that said which problem was detected. And, of course, there needs to be a way for the driver to reset the warning signal without taking the car into the dealer after they tighten their gas cap. A safety override could keep the driver from resetting the warning for a serious problem.

To summarize: warning signals need to be seen and/or heard, but shouldn't cry wolf. To be useful, they have to give enough information for the person being warned to know what's wrong and what they should do about it. Warning signals should go away when the problem has been solved.

Prototypes

An important part of developing any product or interface is prototyping.

There are different types and levels of prototypes. Visual prototypes include drawings or sketches on paper, on a computer screen, or on a whiteboard or blackboard.

Interactive prototypes, whether in hardware or software, are important for design and testing. Software interface prototypes, of course, should be on a computer. Though not necessarily in the final program, they must simulate how the controls will work. Even hardware interface prototypes can be initially prototyped on the computer, and a lot of testing and refinement can be done with this simulation. But eventually, a physical hardware prototype will have to be made to test the "feel" of the controls.

Programming languages are great for onscreen prototypes, but take a few years to master. Tools like HyperCard®, Director® and similar programs allow nonprogrammers to quickly and easily whip up useful early-stage prototypes.

Prototypes should first be tested or reviewed by other team members or local experts and revised, then tested on typical customers. And revised again.

Good Questions to Ask

When you want to evaluate an aspect (or object) of the interface, these are good questions to ask yourself:

- Can the customer tell what actions are possible?
- Is it instantly clear which parts move and which don't?
- Is it obvious where you grab or interact with the object?
- Is it obvious what kind of movement is possible (push, pull, rotate, touch, etc.)?
- Can the customer determine the mapping of the controls to resulting action?
- What are the physical characteristics of the movement? How much force do you need? How much is too much? How do you know when you've moved it far enough—or too far?
- Can the customer perform the needed action?
- Can the customer tell if the system is in the desired state?
- Can the customer tell what state the system is in?

Designing to Minimize Errors

Errors by designers and by end users will never be eliminated completely. At least not until humans are eliminated completely from the whole process. But there are things we can do to make errors less frequent and less painful.

1. Understand the causes of error and design to minimize those causes.

2. Allow as many "undos" as possible.

3. Make non-undoable things difficult to do.

4. Give clear, non-annoying warnings.

5. Change your attitude toward errors. Don't think of a user's error as an unfortunate end unto itself. Think of it as an approximation—a step on the way to the solution. Work from that approximation to help the user get to the right place.

> "Then there is technology, the excesses of scientists who learn how to make things much faster than we can learn what to do with them."
> **— Thornton Wilder**

Philosophy

"The point of philosophy is to start
with something so simple as not to
seem worth stating, and to end
with something so paradoxical that
no one will believe it."

— Bertrand Russell

This chapter covers the things that come up in conversations about writing that don't fit anywhere else. These deal with writing in general; not specifically technical or UnTechnical writing. The information applies to all types of writing. Honest.

Now

The Romance of Writing

There's something romantic about being a writer. When you meet people and they ask what you do, you say you're a writer—and making a living at it—and you can hold your own with the "professionals" as well as the "artists." Being a writer (though few others realize this) means that you can sit alone in a room—something many people cannot do for long—and stare at a blank page or screen, and create. Whether you're explaining something or telling a story, it's creating. And that's the real thrill: creation.

I love being a writer. I love having written. But the reality of it is that the actual writing is plain hard work. And it can be boring and even painful—sitting alone for hours, forced to endure your own company and deal with your inner voices. Many famous writers have used this boredom and pain as an excuse to indulge in alcohol or other inebriants. How romantic. Do you have to live high drama and tragedy to write high drama and tragedy? Do you have to be a computer to write about computers? It might help, but I prefer to lean on imagination and research.

The Fun Part of Writing

For me, the fun part of writing is the beginning. Coming up with an initial idea, point of view or starting point. Letting the possibilities fly. Outlining, planning, organizing the possibilities until there is a blueprint of the document or story to come. It's like the beginning of a relationship. Everything's new and exciting. You're exploring new territory, making discoveries and opening yourself up.

Of course, after the newness wears off, a good relationship is still good, but a little less exciting. And it takes more work, thought and compromise to sustain.[45]

So it is with writing. For me, once the beginning is done, the outline and plan completed, the rest—cranking out the first draft page-by-

---

[45] Note to my wife: But it's all worth it, dear. Really.

page and chapter-by-chapter—is just plain hard work. Sometimes I almost have to chain myself to the desk to keep myself working. I'd rather be beginning some new project. But how you deal with the hard work and how you complete your projects—one of the most difficult things in the world to do when you're alone and there's nobody cracking the whip—is a measure of character and professionalism.[46]

Once the first draft is done, and it's time to refine and polish, then it gets fun again. Not as much fun as the beginning, but far better than that first draft.

I know writers who are the opposite. They suffer through the beginnings, relish cranking out pages, and dislike rewrites. I read a screenwriting book that said that the initial outlining and working out the plot and interwoven underlying bases of the story is the hard part, and writing the actual scenes is the fun part. I sure didn't write that book.

Creativity

How many times have you heard someone, whether in person or in an interview, ask a writer, filmmaker or other creative person, "Where do you get your ideas?"

My reaction is always, "You're asking the wrong question!"

Ideas are easy. They happen all the time. Constantly. So much so that it's an annoyance. What I want to hear from successful creative people is: "How do you stop the ideas from coming so you can finish one before the next one comes along?" Or, "How do you judge which of them are the best?" Or, "How do know which idea you want to spend the next six months to two years on to complete it?"

Ideas are easy. They're a dime a dozen. Good ideas are another matter, of course. They're maybe two dimes a dozen.

What is valuable is the ability to take an idea and turn it into reality. This is worth something because it takes a lot of time, energy and usually, money. An idea for a novel or a computer game or a movie can take a split second, where your brain makes a new comparison or connection between existing ideas or concepts. But writing that novel takes months or years of research and work; creating that computer game

---

[46] *And even more importantly, if you don't finish you don't get paid.*

takes a team of 5 to 25 people, a year or two and a couple of million dollars; and creating that movie takes nearly a hundred people and many millions of dollars. In comparison, having the idea, with no investment, no risk, is far less valuable.

To the point: ideas are good, but doing something with them is far, far better.

I used to have trouble with creativity. I had lots of ideas, constantly. But I never did anything with them, and the creativity didn't impact my life at all.

Many years ago, I read a book by Roger von Oeck, called *A Whack on the Side of the Head*. It's a good, very enjoyable book that very clearly and humorously answers the question, "Where do you get your ideas?" It showed all the tricks, techniques and mental twists and gymnastics that help you look at things in new ways and generate lots of ideas. If you don't have a constant flow of ideas overwhelming you on a regular basis, I highly recommend this book. But while I did enjoy the experience of reading the book, for the most part, I didn't learn anything new. I pretty much used all the techniques already, and reading it hadn't really accomplished anything.

Then he came out with a follow-up book, *A Kick in the Seat of the Pants*. This book, while also enjoyable, was the kick in the seat of the pants that I needed. It went beyond generating ideas to actually doing something with them. And it's simple—in concept, at least.

Basically, he breaks the creative process down into four stages. In each stage you play a different role. The roles and stages are:

Explorer—look around, dig up possibilities.

Artist—work with the possibilities to create ideas.

Judge—look at the ideas and decide which one(s) to pursue.

Warrior—dedicate yourself to the idea the Judge selected and go to war, if need be, to finish it.

And finishing it, whatever it is, is the most difficult and most rewarding thing we can do. And when "it" is your own idea, it's even more difficult and rewarding.

I highly recommend this book to anyone, and have given away many copies as gifts.

What does creativity have to do with technical and UnTechnical writing and writers? Everything. If you solve problems and make decisions, you'll do a better job if you do it creatively. I don't suggest you make up the facts (no matter how creatively), but find creative ways to explain or convey them. There may be technical writers who don't consider themselves creative. But every one I've met does. And they not only put their creative energy into their technical writing, but usually have some other writing in the works, if only to keep their creativity growing.

Decisions

Writing is the ultimate decision-making experience. Every paragraph, every sentence, every word is a decision. Tone and point of view are decisions that affect all the other decisions.

And as stated before, writer's block is nothing but a failure to make a decision.

You as a writer are capable of and well-practiced at making decisions in your writing. And there's nothing stopping you from carrying that decision expertise over to the non-writing part of your life.

Pep talk's over. Get back to work.

The User

You may notice that I generally use the term *customer* or *reader* for the person who purchases and uses the products we write about. The term "user" is generally accepted, and I find myself using it sometimes, but it still bugs me. It has often been said that the only two industries where the customers are called users are computers and illegal drugs.

Mac or Windows?

Unless you're working at a humongous company with workstations on a minicomputer or mainframe, you'll be working on a Mac or PC.

There are those who swear by one or the other, and are on personal missions to convince others that their preference is the only computer system to use.

But really, they both work fine, and these days you can move data between them easily without files getting corrupted or strange symbols appearing. Excellent graphics and word-processing programs are available for both computers that allow seamless transfer of data from one machine to another.

So the answer to the Mac or Windows question, is "either." Or both. Or whichever you like, whichever you can afford, whichever is placed in front of you when you get a job, whichever runs the software you want, whichever your clients like. Computers are just tools to get your work done. You may prefer one brand of hammer over another—it may fit your hand better or absorb the shock a little more—but both hammers drive in those nails.

Play it safe and be prepared to work on either type of computer. It's not difficult. The interfaces differ, but conceptually (especially for an intelligent techie like you), they do the same things and pretty much run the same programs. You must know people with each type of computer that you can use for a few hours to get used to them.

Some people have political or personal reasons to like or not to like one type of computer or the other. Fine. Be that way. They both work. They both crash. They both have their problems. Yes, Windows '95 is pretty much like Mac '84, but the newest Macs are also pretty much like Mac '84.

My ambivalence (possibly rare in this matter) stems partly from the fact that when I was in school, there were no Macs or PCs to get familiar with and loyal to. Then, in the many years I worked for a software company, I had both. We created games for both Macs and PCs, so I would run the software under development (usually still buggy and crashy) on one computer and write the manual on the other. When it was time to work on the other version of the program, I swapped the machines.

There are more important things to argue about than which computer to use.

Credit—and Blame—Where It's Due

The general trend in technical manuals is to mention the authors' names on the credit page, buried in with a hundred other names, and not on the title page. Authors' names belong on the title page.

There's no thrill for an author like seeing a book or other document, finished, bound and sporting their name. Aside from the adulation of millions of readers and boatloads of money (neither of which are likely to happen), this is what makes the writer feel that all that work, all those hours alone with the keyboard, and all those review and editing sessions were worth it.

Look at this as a perk. A lot of writers will take one job over another, even at slightly lower pay, if they know they'll get credit and their name prominently placed on the title page. This goes for contractors as well as employees. A bound book with your name on the title page is a great résumé piece.

Also look at this as a guarantee of good work. If your name's on it, you're responsible for it. You get the credit if it's good and the blame if it's bad. If you know your name's going on it, you're going to put in extra effort to make sure that that document is the best you can make it.

And on the subject of credit, if you can, spread it around. The administrative assistant who tested your tutorial, the tester who corrected some important content issues and others who contributed to the document all deserve a mention somewhere on the credit page, possibly in the Special Thanks To: section.

Manual v. a Stand-Alone Book

Since this book was written to stand on its own, and doesn't come with a product, I will assume that you are reading it because you want to. For whatever reason: you want to hear what I have to say or you want to learn something about UnTechnical writing, or want to prove that I'm full of bull[47] or whatever ... but you chose this book, and you can read it or not at your whim.

---

[47] *And you're welcome to it. Please send me any mistakes, omissions, problems, etc. Some call it criticism, I call it editing.*

On the other hand, the person who buys a VCR or a computer program chose the VCR or computer program. They didn't choose the manual, yet they may have to read it. In the mind of the new owner, the manual is a necessary evil. What they want to do is use their new whatever it is, not read the manual.

Therefore, while I can torture you endlessly in a book like this with my ramblings and personal philosophy, you have to make the manual as easy and painless as possible for the reader, in writing and layout. It is up to you to trick, tease or bribe the new owner into looking through the manual and getting the most out of it.

For the stand-alone book, you don't have to bribe or trick. Just make it useful and entertaining.

Poetry, Philosophy and Technical Writing

When I want to wrestle over the meanings of words and interpret the author's meaning, I read poetry.

When I want to spend time pondering deep thoughts and delve into heavy, meaningful paragraphs, attempting to pull some relevance to my life out of them, I read philosophy.

When I want to find out as quickly and simply as possible how to do something—and then go do it—I read technical writing.

Now to confuse the matter and contradict myself: if you can slip a little poetry and philosophy into your tech writing without slowing things down or getting sidetracked, then by all means, do so.

A Mission or a Job

Looking back on all the jobs I've had—and I've had jobs in a number of industries at levels ranging from burger cook and factory-worker to management to owning my own businesses—the times I remember as being the most fun or "the good old days" were the times when I felt like I was on a mission. It wasn't the amount of money I made (though I am fond of big, fat paychecks), the job title or the number of reports (not that I have any aversions to holding the reins of power) or the accolades of a job well done (insert your own wisecrack here) that brought the most enjoyment.

It was the work itself—but more than that it was working as a team with a few others toward a common, nearly impossible goal. It's like the scene in the movie *The Last Starfighter,* when the big brown alien says something like, "I've always wanted to fight an impossible battle against insurmountable odds." We laugh at that line, but it's true. And that team feeling can be especially enticing and rewarding to writers, who spend so much of their work time alone.

I know I did better work and had more fun when I felt like I was part of a team, on a mission. Any time I was just working for the sake of a paycheck, and didn't really care about what the work was and didn't have a tight team feeling, it was just selling part of my life instead of living it.

So what's in this for you? Should you rush out in search of a mission? Should you quit your job and find a goal?

Probably not.

But look around you. You may be on a mission, and not realize it yet. These are the good old days you'll look back on with pride. You might as well enjoy it now.

> "There's only one way to work—like hell."
>
> **— Bette Davis**

> "Run, if you like, but try to keep your breath;
> Work like a man, but don't be worked to death."
>
> **— Oliver Wendell Holmes**

Later

The World Will Be a Better Place....

This may seem like sacrilege to many technical writers, but I believe that the world would be a far better place if there were less technical writing for the consumer market.

Don't worry, there will always be a need for technical writing and technical writers—to provide information to technical people with technical jobs. But for the consumer market, we could do with much less.

How? Better product design. And who will make sure the products are designed better? The writer/designers that used to think of themselves as technical writers. Yes, you. You've got the communications skills and the technical background. Use it.

A question for you experienced tech writers: how many times have you spent hours trying to figure out how to explain to the reader how to do something, and thought, if only it were designed this way it would be easy to explain? It happens all the time.

If you're lucky enough to be working for or with a company that gets you involved in the design stages of a project, you've probably had an occasional opportunity to make suggestions that not only make the product easier to describe (easier for you), but make the product better because it's easier for the customer to understand.

In my humble opinion, writers need to stop thinking of themselves as the people who write the manuals, and start thinking of themselves as communicators of information. The information may be in a manual, it may be in an onscreen help window, it may be in the design and interface itself. Knowing where and how to present information should be part of the writer's job, part of the writer's expertise.

The Future of Technical Documentation

For both financial and technical reasons, technical and UnTechnical documentation—at least for software—is heading away from being printed on paper and toward integration with the product itself.

The financial reasons are that paper is getting more and more expensive, while memory, RAM and disk, is getting cheaper. The technical reasons are that products—both hardware and software—are getting smart enough to contain their own help systems. And we also hope that the products will be designed well enough that they'll be able to be used intuitively, and not need as much documentation. (We'll see.)

Instruction videos are another direction for technical materials that will be around for a while. While it is fairly cheap to manufacture a VHS tape (and DVD will be cheaper), video, too, will head toward integration with the product whenever the product has a display screen (for instance, like a television, VCR or cellular phone).

As the rate of data transfer speeds up, documentation and instructions will more and more go online, as opposed to just onscreen. A master copy of the docs, tutorials, instruction video or whatever will reside on the net, and the product will pull it down and display it as needed. This master copy will always be up-to-date.

Of course, any product that uses online help will have to be hooked to the phone lines or cable or some other connection to the net. This may seem far-fetched now, but it'll happen. Houses will be networks. (A number of companies are developing home networks, including some based on Microsoft's Windows CE.) Appliances will talk to each other for timing, coordination and energy efficiency, and to the factory to get updates, help screens and warm fuzzy messages and helpful hints for the owner.

It will be a while before technical and instructional books and manuals are totally gone. They just work better much of the time, and customers are still much more comfortable with them. Then again, technical information is growing and changing at breakneck speed. Printed material is stagnant, and can quickly become outdated. Over the next generation or two, printed technical information will become rare.

In the meantime, as both hardware and software become more self-contained, printed manuals—or the parts of the manuals that give basic operating instructions—generally become smaller. This doesn't mean that there's less for the writer to do.

There's still:

- Online or onscreen help systems,
- Background and added-value printed material, and
- Product and interface design.

Remember, your job is to get the needed information into the customer's hands and mind as quickly and easily as possible. Don't get left behind by thinking that your method of communication must be through a printed manual.

Continued Research

Maybe I've been working in software too long, but I think of this book as version 1.0. I've spent years gathering information and figuring things out, and I wrote this book as accurately and to the best of my ability at this time in order to share information with other writers to try to help make their work and lives better.

But things change with time and further experience, and I fully expect to have a version 2.0, and possibly 3.0 of this book over the coming years.

If you want to contribute your knowledge or experience to the general writing population without writing and publishing an entire book, I'll be glad to help act as a conduit. I'll consider anything sent to me, from tools and techniques to anecdotes and horror stories as possible additions to future versions of this book. In return for your shared wisdom, I offer credit—to you, but, sorry, not to your bank account.

Also, I'd like general feedback on this book from you readers and writers, including:

- What have I missed?
- Where do you think I'm wrong?
- With what do you agree/disagree?

See Contact Information at the end of this book for ways to help/inform/insult me via email, snailmail or my website.

Exhibits

"Well, for those who like that sort
of thing I should think that is just
about the sort of thing they
would like."

— Abraham Lincoln

This section of the book has sample forms, lists and worksheets that may
be useful to you in your writing.

About the Exhibits

Some of the sample documents here were described or at least mentioned in the text of this book. Others weren't mentioned, but are here because they're either useful or potentially interesting.

Use these sample documents as a guide, not as gospel. Copy and modify them to your needs.

Generic Deliverables List

Here's what I use as a generic deliverables list for a computer game. It includes all the tasks that the writers do over the duration of a project.

I keep this list in spreadsheet format for easy recalculations, and also so it could later be expanded into the Project Breakdown document.

Before new projects are fully funded, producers come to me for writing resource and cost information. They explain the project as well as they can, and between us we come up with a time and cost estimate. I use this list to both help me be complete in assigning time for all the tasks, but also to remind the producer of all the tasks that writers could be—and usually were—involved with.

The producer's estimate was included in the proposed project budget. If the project received the go-ahead, then it was time to do a detailed project breakdown.

This first estimate is always rough, since it is done before the project has been designed. It is usually updated at least a couple of times as the project becomes more concrete.

Online and Onscreen

There was a time when online help in the software world meant that it was on the disk and you read it on the screen. Nowadays, online means that the computer is connected to some remote source of information on the line, over the phone, through the cable, via satellite, etc.

In this book, help that resides on the local disk or local computer and is displayed on the monitor is "onscreen." "Online" is reserved for accessing remote information via "the line."

Here are the samples, one blank, one filled in for a theoretical project, Personal Newspaper (which is used as the sample project in the book *The User Manual Manual,* also from UnTechnical Press).

Generic Writing Deliverables

Deliverable	Description/Quantity	Est. Hrs.	Est. Wks.	Notes
Research				
Design Doc Work				
Prototype Work				
Manual				
Quick-Start Guide				
Screen Text				
Help System				
Readme				
Teacher's Guide				
Internationalization				
Totals				

Research is a catchall term covering various types of prep-work. Since the company I was at made simulation games, research generally involved gathering information on the system being simulated, as an aid to the designer. This could also involve researching similar products and putting together reports on various features that might need to be included or improved.

Design Doc Work was any time spent on design documents. This could be a simple edit for clarity—to make sure the designer's points were being communicated well—or the actual writing, depending on the skills of the writer and the various others on the design team.

Prototype Work involves everything from supplying simple screen text for a prototype, to documenting or designing prototypes, to actually building a prototype on paper, as a model, or in HyperCard, Director or another computer program.

Generic Writing Deliverables—Sample Project (Personal Newspaper)

Deliverable	Description/Quantity	Est. Hrs.	Est. Wks.	Notes
Research	Library and Internet research, 2—3 field interviews w/experts, compiling information	120	3	Interviews with at least one reporter and one editor.
Design Doc Work	Editing only	20	.5	Designer will write doc, only needs editing.
Prototype Work	10 Sample stories incorporating the designer's name variable system to test the database	20	.5	At least two of the stories have to be long enough to test multi-page print.
Manual	Will only need about 60 pp. Geared for parents	160	4	Includes editing, testing and layout edits.
Quick-Start Guide	8 pp. incl. cover. Fits into CD jewelcase	40	1	Includes editing, testing and layout edits.
Screen Text	Mostly editing, some naming	20	.5	
Help System	Standard Windows Help System, context-sensitive, based on manual, with extra "How do I ..." section	80	2	Includes editing, testing and debugging.
Readme	The usual	10	.25	
Teacher's Guide	Will use standard company teacher's guide template	160	4	Includes editing, testing and layout edits.
Internationalization	No I18N planned unless US version sells well. Just organize, prepare and store materials just in case.	10	.25	Since this project is so text heavy, I18N will be very expensive and take a lot of time.
Totals		640	16	

Manual includes writing, rewriting, customer testing, editing and shepherding the manual through layout.

Quick-Start Guide includes writing, rewriting, customer testing, editing and shepherding various short documents and look-up cards through layout.

Screen Text is writing or editing the basic interface text: menus, buttons, etc., as well as back story, dialog and explanations of program elements, and even an interactive tutorial. This includes every bit of text that the customer will see on the screen that isn't contained in a help system.

Help System is the standard or custom onscreen help.

Readme is the traditional last-minute information that accompanies most software. It contains information important to the customer that wasn't ready or wasn't known in time to be included in the regular docs. Typical contents include extra installation instructions and recently found incompatibilities with various hardware and software.

Teacher's Guide is the writing, editing, classroom testing and laying out of a teacher's guide. This is only done on projects that are likely to be used in schools.

Internationalization includes preparing, rewriting, and generally communicating with translators.

Notes

- This list and the comments are fairly software-specific, but you can adapt it to whatever type of project you work on.

- Depending on your company and its business staff, you may be able to make your time estimates in weeks, or you may have to break it down to the hour.

- If you are a contractor, you can use a list like this to help you estimate time and set your rate.

- Since this form was primarily for budgeting, the time entered was the total time (time = money) it would take to complete the task, no matter how many different people would actually perform the work. Later, in the project breakdown, the different tasks could be assigned to different people.

- The tasks won't be completed sequentially, one after the other. There could be months between research and the next tasks. Other tasks will overlap, and the writer(s) will jump from task to task (or project to project) as the project progresses and the information becomes available. The times listed are the actual times to complete the task, not necessarily the duration of the work. A task that is listed to take one week could take place over a two- or three- or six-week period. But it would only take 40 hours during that period. Once again, this is because this information is for budgeting.

Suggestions

- Modify the list for your own generic project, then customize it for each individual project.
- Track actuals vs. estimates so you can update your estimates and be more accurate next time.
- Send me samples of your generic deliverables lists for your types of projects, and I'll include them in the next edition of this book or make them available on the UnTechnical Press website.

Project Breakdown

This form is more useful for writing managers who are coordinating the work of multiple writers on multiple projects than for individual writers, but it may be interesting for others as well.

The project breakdown is a tool for breaking down all the deliverables by time and person.

This really has to be a spreadsheet, with the different sections tied together. I use a spreadsheet that allows multiple sheets within a single document that can easily be tied together, but it can all be on a single sheet if you don't mind a lot of scrolling.

There are two types of sheets: Project sheets and the Summary sheet.

Project sheets contain:

- **The deliverables list.**

  This list is customized to the project. You may want to break it down further. For instance, you could divide the manual into separate writing, testing and editing tasks, but I've kept it simple here to keep the example clear.

  In addition to the time estimates in hours and weeks, this section of the summary includes a place for due dates (1st draft, final draft, to graphic arts and to manufacturing) for each deliverable.

  It also has Scheduled (Sched.) and Remaining (Rem.) columns.

  These are just to help you keep track of what you have and haven't scheduled. As you assign the tasks or portions of tasks in the breakdown below, put the portion of the task that you have assigned and scheduled in this column. It saves you the task of remembering what you did and didn't assign, and saves the time of adding up hours.

- **A breakdown of who does what when.**

  Here is where you do your actual planning. Using the due dates and estimated times, assign the various tasks to the people who are going to do them. The form allows assigning work to staff writers, temps and contractors. All tasks on a whole project may be

```
Project Breakdown.xls                                                    _ □ X
```

	A	B	C	D	E	F	G	H	I	J
1	**Project Breakdown--Project 1**									
2	Start:	Jan 1								
3	Writing Completion Date:	June 30								
4										
5	**Deliverables**	**Description/Quantity**	**Est. Hrs.**	**Est. Wks**	**1st Draft**	**Final Draft**	**to GA**	**to Mfg.**	**Sched.**	**Rem.**
6	Research	Library, Internet	160	4	Jan 21	Jan 31	n/a	n/a	160	0
7	Design Doc Work	Editing Only	20	0.5	n/a	Feb 25	n/a	n/a	20	0
8	Prototype Work	Text for prototype	40	1	n/a	Mar 15	n/a	n/a	40	0
9	Screen Text	Write, edit interface text	60	1.5	Apr 5	15-Apr	n/a	n/a	60	0
10	Onscreen Tutorial	Pop-up dialogs	40	1	May 5	May 10	n/a	n/a	40	0
11	Manual Writing		120	3	May 15	May 30	n/a	n/a	120	0
12	Manual Editing	Incl. editing layout	60	1.5	n/a	Jun 15	Jun 15	Jun 30	60	0
13	Help System	From manual, w extra "how do I .." section	80	2	Jun 20	Jun 30	n/a	n/a	80	0
14	Quick Start Guide (write)	4 pp., fit into CD jewel case, incl. testing	30	0.75	Jun 10	Jun 15	n/a	n/a	30	0
15	Quick Start Guide (edit)	Incl. editing layout	10	0.25	Jun 16	Jun 20	Jun 20	Jun 30	10	0
16	Read Me	The usual	10	0.25	Jun 25	Jun 30	n/a	n/a	10	0
17	Internationalization	Organize and prep for xlation	20	0.5	Jun 15	Jun 30	n/a	n/a	20	0
18										
19										
20	**Totals**		**650**	**16.25**						
21										
22	**People/Task Breakdown**									
23		Task	Jan	Feb	Mar	Apr		May	Jun	Totals
24	Writer 1	Research	160							160
25		Design Doc		20						20
26		Prototype			40					40
27		Screen text			30	30				60
28		Onscreen tutorial						40		40
29		Manual Writing						60	60	120
30		Quick Start Guide							30	30
31		Readme							10	10
32		I18N							20	20
33	**Writer 1 Total**		**160**	**20**	**70**	**90**		**100**	**60**	**500**
34										
35	Writer 2	Manual Editing							60	60
36		Quick Start Guide (Edit)							10	10
37										0
38										0
39										0
40	**Writer 2 Total**		**0**	**0**	**0**	**0**		**0**	**70**	**70**
41										
42	Writer 3	Help System							80	80
43										0
44										0
45										0
46	**Writer 3 Total**		**0**	**0**	**0**	**0**		**0**	**80**	**80**
47										

```
Summary \ Proj 1 / Proj 2 / Proj 3 / Proj 4
```

done by a single person, or many people may share a single task. It all depends on who's available, who has the best skills for the task, and your management style.

This project is primarily Writer 1's responsibility, but I'm utilizing Writer 2's editing skills and Writer 3's experience with online help systems.

The Summary sheet contains:

- **Headcount Summary**

  This is a month-by-month display of how many writers you will need to complete all the projects as scheduled on the project sheets. This is your ultimate bottom line for the writing group, and your boss will be very interested in it.

  Things to look out for are peaks or valleys where you need considerably more or fewer writers than you have. When you see peaks, you need to either reschedule work (when possible) to available writers in slower months (valleys), or be prepared with temps or contractors at the proper time. If your required headcount is consistently higher than your staff, you could consider increasing your permanent staff.

  In a perfect world, the sample below would show the required count the same each month—however many writers there are on the team. In the real world, it doesn't work like that. On paper, it'll look like your staff has nothing to do some months, and you're a cruel slave driver the rest.

- **Headcount Summary by Project**

  This is a project-by-project monthly breakdown. It lets you know at a glance what time and resources are scheduled for each project. The individual project producers will be interested in this.

- **Time Load by Person**

  This shows the total hours a month each writer has been scheduled, with an additional breakdown by project by month. This is really of most concern to you. You've scheduled your staff on multiple projects, one at a time, and here's the result.

  It's amazing how quickly and easily you can schedule people to work 80 or more hours a week. But unless you want shoddy work or an angry mob, you've got to go back to the project sheets and redistribute the work to a more manageable load per person.

  Think of this sample as a first draft. You make your best guess at who needs to do what when. It's close, but not good enough. Go

Writing Group--Project Breakdown--Summary Sheet

	Jan	Feb	Mar	Apr	May	Jun	Totals
Total Headcount	3.34	2.31	2.31	1.97	3.16	3.97	17.06
Headcount by Project							
Project 1	1.00	0.13	0.44	0.56	0.63	1.31	4.06
Project 2	1.19	0.31	0.44	0.31	0.69	1.56	4.50
Project 3	1.06	1.75	1.25	0.94	1.63	0.94	7.56
Project 4--Misc.	0.09	0.13	0.19	0.16	0.22	0.16	0.94
Time Load by Person							
Writer 1	180	85	120	180	165	70	800
Project 1	160	20	70	90	100	60	500
Project 2	0	0	40	30	0	0	70
Project 3	10	60	0	40	40	0	150
Project 4 Misc.	10	5	10	20	25	10	80
Writer 2	165	120	110	75	145	290	905
Project 1	0	0	0	0	0	70	70
Project 2	160	50	30	20	80	190	530
Project 3	0	60	80	50	60	30	280
Project 4--Misc.	5	10	0	5	5	0	25
Writer 3	190	165	140	60	195	275	1025
Project 1	0	0	0	0	0	80	80
Project 2	30	0	0	0	30	60	120
Project 3	160	160	120	60	160	120	780
Project 4--Misc.	0	5	20	0	5	15	45

back to the project sheets and see what tasks you can move around to a different month or a different writer, and try to get all your writers' projected monthly hours close to 160.

Whatever you do, don't schedule less hours than a task really needs to be done well. If, after all the moves you can make, your total headcount is too high, start looking for temps or contractors to fill in during the busiest times.

Notes

- As stated before, this is more of a writing manager's tool than a writing tool. If you don't need it, don't worry about it.

- All numbers on the Summary sheet are calculated directly or indirectly from the numbers on the Project sheets. Never enter numbers on the Summary sheet.

- These printed samples only cover three writers and four projects over six months (small, simple projects), so they can be shown on a printed page. If this is all you have to worry about, then you might not need to bother with linking everything together in a spreadsheet. But if you have 10 or more different projects going at once, and a combination of staff, temporary and contract writers, each with their own skills and talents to make the most of, then the time it takes to do a spreadsheet like this is well worth it.

- The Misc. project covers all the various requests for writing or editing smaller projects ranging from newsletters to memos. These things don't each need a full project sheet, but they do take time and need to be scheduled. Lumping them into a single project keeps them from surprising you and draining your resources.

Suggestions

- Put real names for projects and people into the sheets.

- In the Project sheet template, you can add and link slots for temporary writers and contractors.

- Run the sheet out to at least two years.

- If you're a spreadsheet wizard, then you can really get fancy. For starters, you could set up warnings in the project sheets that tell you when you've overscheduled a person. Beyond that, who knows?

Writing Request Form

When you've got a large, extended project, it can be arranged through meetings with the producer and the use of the previously described Project Breakdown sheet. But if you're not scheduling multiple writers on multiple projects, that form is overkill. Plus, there are many smaller projects that, though they don't need detailed planning, still need to be described and scheduled *in writing* (and grouped together in the Misc. project) to avoid inevitable problems.

This is a simple form that works as an information checklist and approval form for simple jobs. It can be distributed in hard copy or electronically, but I found that the usual drill was when someone needed some writing, they'd come to me, and I'd ask the questions and let them go. Then I'd fill out most of the form and dig out my Project Breakdown sheet and see who had the time and skills to fulfil the request. After checking writer availability, I'd fill in the bottom of the form letting them know exactly when it'd be ready.

The top part of the form is General Information—who wants what, and where does the money come from (account code).

The second part of the form is contact information. It doesn't have to be the whole development team, just the names and numbers of the people that might have answers you'll need to evaluate how long the project will take and make your promise (or not) to get it done by the requested time.

Next is the task list, a short breakdown of what's needed when, and where to send it.

At the bottom is a place for writer comments (it will/won't/can't be done, or will be done early, etc.), and approvals. Approvals are a pain and they make you feel like a paper-pusher, but it's usually worth it to make sure you're getting paid before you do the work.

Yes, forms like this are annoying and officious, but they can save you time, money and hassle in the long run. Getting everything that you can in writing is common sense.

Writing Request

General Information

Project Name/Title: _____   Account Code: _____

Requested By: _____   Request Date: _____

Brief Project Description: _____

_____

Contact Information		
Name	Title	Phone/email

Task List

#	Task	Description	Due Dates		Deliver To	Format
			1st Draft	Final		
1						
2						
3						
4						

Comments

Writer Comments _____

_____

Approval and Acceptance

Requestor   For Writing Group

Name, Date _____   Name, Date _____

Title _____   Title _____

Project Contact List

During the course of a long writing assignment, chances are you'll need to contact the various people on your project. It's nice to have a project personnel list by your phone or computer so you can easily contact anyone you need.

Be sure to include people from management, marketing, engineering, testing, and anyone else you might possibly need to contact. Make it a table and expand it as large as you need.

The Notes section is for best times to contact, their location in the office, or anything else that occurs to you.

Project Contact List

Name	Title	Phone	Email	Notes

Customer Fact Sheet

This is a very optional form for those who want or need to be reminded who their customer is and who they're writing for. Put a customer fact sheet up on the wall near your computer and look at it every so often as you write.

The boxes at the bottom of the page are for pictures of typical customers. If you can't get names and pictures of actual customers, cut pictures out of magazines.

Customer Fact Sheet

Project:	
Average age:	
Age distribution:	
Gender distribution:	
Technical expertise level:	
Education level distribution:	
% repeat customers:	
Professional field:	
Other special attributes:	

Name

Name

Name

Name

Checklists

Checklists are a way to keep things from falling through the cracks and staying focused.

The first one is a generic writing-project checklist. It's modeled on writing for a computer game, but it will have a lot in common with whatever writing you're doing. In any event, you'll need to customize it to your needs for almost any project you do.

The other checklists are more detailed step-by-step task lists for individual parts of a project. Again, these are based on software in general and games in particular, so be prepared to modify them for your own needs.

The checklists shown here are:

- Writer's Overall Project Checklist—this is for a staff writer who is involved in a project from the beginning to the end. A contractor or writer who is brought in only at the end of a project won't need this list.

- Manual Writing Checklist—this is a basic list of the steps a writer goes through on a large manual on a large project. Lots of checks and rechecks.

- Quick-Start Guide Checklist—this is for either a supplement to a manual or a very short manual.

- Editing Checklist—this list details the whole editing process.

Writer's Overall Project Checklist

Before the Project Begins

❑ Project Approval/Request Forms

❑ Project contacts and team members

❑ Location of project network storage

    ❑ Project Definitions

    ❑ Definition/description of the product

    ❑ Intended purpose (pure fun, pure ed., home learning, school learning, home creativity, combinations)

    ❑ Intended audience (Customer Fact Sheet)

    ❑ Projected platforms and languages

    ❑ Projected schedule and due dates

Initial Design Phase

❑ Estimate project documentation needs

    ❑ Estimate need for printed documentation

    ❑ Presence and size of onscreen or online docs

        ❑ Will they exist?

        ❑ How extensive? (small pop-up help, complete onscreen documentation?)

        ❑ What format(s)? (web- or disk-based, platforms, etc.)

    ❑ Project needs for screen text (other than help)

❑ Help create/maintain Initial Design Doc

❑ Project time/personpower for printed docs, onscreen docs and screen text

❑ Help prepare for Initial Design review

❑ Help prepare presentation for Management Team review

Final Design Phase

❑ Help create/maintain Final Design Document

❑ Push for interface design finalization: menus, window names, all dialogs, etc.

❑ Prepare preliminary listing of all onscreen text (push to have all screen text in separate, editable files)

❑ Prepare outlines for printed docs and onscreen help(s)

❑ Help prepare for Final Design review

❑ Help prepare presentation for Management Team review

Production Phase

❑ Write/edit/test reference section

❑ Write and test tutorial

❑ Finalize tutorial

❑ Manual introduction, front matter and back matter

❑ Write/edit/test onscreen docs/help

❑ Write/edit/test screen text

❑ Write package copy or prepare relevant information for package copywriter

❑ Supply preliminary copy of printed docs to Internationalization

❑ Supply preliminary copy of onscreen docs/help to Internationalization

❑ Supply preliminary copy of screen text to Internationalization

Final Checklists for the End of the Production Phase

Printed Docs

❑ Edited, spell-checked, reedited

❑ Checked for proper reading level

❑ Proper format for turnover to GA

❑ Check smart quotes/em-dashes/symbols, etc.

❑ Graphics prepared in proper format

❑ Graphics list

- ❏ Retain copy of all graphics
- ❏ Turn over to GA
- ❏ Final text provided to Internationalization
- ❏ Post-layout edits
- ❏ Create index, TOC, etc.

Onscreen Docs

- ❏ Written, edited, tested
- ❏ Context IDs to programmers
- ❏ Graphics taken, in proper format and resolution
- ❏ Help file compiled and tested, including context-sensitivity
- ❏ Readme file written and edited
- ❏ Final copies provided to Internationalization

Screen Text

- ❏ Written, edited, tested, reedited, retested
- ❏ Final complete listing of screen text supplied to Internationalization

Cleanup

- ❏ Archiving

Manual Writing Checklist

Phase 1: Preliminaries

❑ Meet with relevant people (group manager, producer, developer) to discuss software's concept, current state, projected evolution and initial schedule

  ❑ Determine audience demographics and tone of manual

  ❑ Discuss possible added-value material (accompanying background essays, research, etc.)

  ❑ Will the manual be converted into onscreen help?

❑ Begin familiarization with the software, with producer, if possible

Phase 2: Initial Steps

❑ Thorough saturation with product

  ❑ Notes on major features

  ❑ Notes on quirks, less obvious bugs

  ❑ Notes on possible interface changes from writer's view

❑ Outline manual; projected page count

  ❑ Possible reevaluation of schedule, preliminary writer's milestones

❑ Write preliminary introduction to help cement the flavor of the project (run it by marketing)

Phase 3: Beginning

❑ Confer with producer about current completion state of software (dig for honest evaluation)

❑ Decide which manual sections can be written in which order, and when screenshots can be taken

❑ Develop outlines for reference and tutorial

❑ Submit outlines to group manager for comments and structural editing

Phase 4: First Draft

❑ Implement changes from structural edit

❑ Write preliminary reference and tutorial

❑ Test tutorial with at least two "typical customers" as well as someone from tech support

❑ Submit documentation to producer, group manager and other team members for content edit

Phase 5: Second Draft

❑ Implement changes from content edit

❑ Supply more precise determination of page count, revised schedule

❑ Discuss scheduling, special needs with graphic arts

❑ Confirm with producer (signed in blood) that no more major software changes will occur (feature freeze)

    ❑ Are all command functions/features implemented?

❑ Determine software's suitability for screenshots

    ❑ Is onscreen text final?

    ❑ Is artwork final?

❑ Update entire document (including introduction) to (theoretically) finished software

❑ Run a copy by the producer, testers and others for additional content check

❑ Conference with producer/developer about final issues

❑ Implement corrections from additional content check and meetings

❑ Submit for line edit

Phase 6: Final(?) Draft

❑ Implement changes from line edit

❑ Take screenshots

❑ Produce graphics list w/comments

❑ Submit revised docs, graphics and list to graphic arts

  ❑ Possibly submit in "chunks" (that is, reference, then tutorial ...)

❑ Beware of ongoing product changes

  ❑ Make list of new edits/revisions, screenshot changes

Phase 7: Preliminary Visual Edits

❑ First look at laid-out document

  ❑ Make sure any special needs, such as title-page graphics, artwork are understood

  ❑ Comment on styles

  ❑ Check that graphics are correct and clear

  ❑ Check that captions, callouts and headings are properly displayed and in the right place

  ❑ Mark index entries

  ❑ Check and edit table of contents

  ❑ Submit last-second product-change information

❑ Second look

  ❑ Edit index entries

  ❑ Check for first-look corrections

  ❑ Submit more last-second product-change information

Phase 8: Visual Edit

❑ Detailed, complete visual edit by a fresh set of eyes (or two)

❑ Review the edits to be sure none were missed

❑ Final edit and proof (once more quickly through the ringer)

❑ All concerned parties (group manager, editor, producer, developer) review material and sign off

Phase 9: Ship It

❑ Check bluelines (if any)

Quick-Start Guide Checklist

Preliminary

❑ Get schedule information, approvals

❑ Determine audience demographics and tone of document

❑ Contacts: Who's on the team? (unless you have this from the Manual Checklist)

Write System Requirements

❑ Mac, Windows, other, multiple

❑ Specific computer models, Power Mac/PC, etc., and exceptions

❑ Hardware requirements

❑ Software requirements (operating system, etc.)

❑ Additional recommendations

Write Installing and Starting Instructions

❑ How to install

❑ Platform-specific instructions

❑ Full or partial installation an option? Pros and cons of each

❑ Starting the product

❑ Registration (electronically or otherwise)

Write Notes

❑ Corrections, additions to manual

❑ Other items of interest

Testing and Finalizing

❑ Test installation instructions (regular testers, tech support and typical customers)

❑ Have everyone concerned check and sign off on System Requirements

❑ Rounds of editing

❑ Layout

❑ Rounds of editing

❑ Final sign-off

❑ Bluelines

Editing Checklist

Stage 1: Outline (Structural Edit)

❏ Examine the document's outline for overall structure, organization and order of information

❏ Look for missing or superfluous subject areas.

Stage 2: First Draft (Content)

For Main Editor:

❏ Suggest changes in tone or slant

❏ Look at readability, structure, focus and logic

❏ Consider moving entire paragraphs around or removing paragraphs altogether

❏ Look at cohesiveness (does one paragraph flow from the preceding one and to the succeeding one?)

❏ Suggest amplifying or discarding ideas

❏ Look at the way ideas are developed, supported and elaborated upon

❏ Pass the document by relevant (if any) personnel (e.g., testers, producers, tech supporters, marketers, etc.)

For Content Editors (producers, testers, programmers, designers, etc.):

❏ Check accuracy of information (especially last-minute changes to the program, window names, etc., that the writer might not know about)

Stage 3: Corrected Manuscript (Line Edit)

❏ Consider construction and phrasing

❏ Check spelling and grammar

❏ Verify the correctness of things like phone numbers, technical specs, legal issues, proper name spellings, etc.

❏ Check sentence rhythm (read aloud.)

Stage 4: Layout or Page Proofs (Visual Edit)

❑ Check corrections against the original edited manuscript (corrected problems often create new ones)

❑ Look at alignment of words and graphics on the page

❑ Check letter and word spacing as well as *leading,* or line spacing

❑ Look at line breaks and page breaks

❑ Eliminate widows and orphans

❑ Check the correctness and clarity of graphics, verifying placement and making sure that each graphic is the right one

❑ Check words within graphics (legible?) and callouts (right one? spelling? initial caps? etc.)

❑ Verify pagination (missing pages?) and the content of chapter identification headers and footers

❑ Check headings: levels, styles, spacing, etc.

❑ Make sure typestyles and fonts are correct

❑ Check lines that end a paragraph for continuity (they can sometimes be "rolled up" and obscured by the program)

❑ Check title pages and chapter heads

❑ Verify table of contents (TOC) and index page accuracy

❑ Make sure you see all pages (front pages and title pages may not have been included with previous versions you edited)

Stage 5: Blueline

❑ Check corrections against the original edited manuscript

❑ Without reading the document, scan every page looking for errors with graphics and text

❑ If time, look over a few random pages more thoroughly, checking headings, sentences and paragraphs, and that text flows from one page to another (If there are many errors on these random pages it probably means you should look the whole thing over again, in detail.)

❑ Re-verify table of contents page accuracy

❑ Re-verify index page accuracy (select a few random ones if it's a large document)

❑ Re-verify pagination

Preparing Draft for Layout Worksheet

This combination worksheet and checklist is for your own use, to help you gather the information you'll need to prepare your draft for hand-off to graphic arts.

Info for You

Document Name: _____ Draft#: _____ Due Date: _____

Deliver electronic versions to: _____

Email: _____ Phone/Ext. _____

Location on Net _____

Deliver printed versions to: _____

Location _____

Electronic Format for text:

      ❑ Mac   ❑ PC

      ❑ Word Processor (_____) ❑ .RTF ❑ Simple Text ❑ Other _____

      ❑ Styles/headings assigned ❑ Styles/heading marked in text [H1]

Electronic Graphics Format(s)

❑ TIF   ❑ PICT   ❑ PCX   ❑ JPG   ❑ BMP   ❑ WMF ❑ Other _____

Info for Graphic Artist

Draft Contains:

      ❑ Number of Tables _____ ❑ Number of Graphics _____

      ❑ Other _____

Printed Draft   Number of Pages _____

Printed Graphics   Number of Pages _____

Info from Graphic Artist

Estimated date/time for pre-edit edit _____

Checklist:

Printed Draft

       ❑ Page numbers

Printed Graphics

       ❑ Page numbers

       ❑ Graphics named, marked up for callouts and captions

Other Printed Material

       ❑ Cover letter/notes

       ❑ Style definitions

       ❑ Graphics list

Electronic Graphics

       ❑ Backed up   Location _____

Sample Content-Editing Cover Letter

When sending out copies of a document for content editing, a cover letter helps to keep what you need and expect absolutely clear.

[Reviewer's name],

Here is a draft of [document name] for your comments. As the project/company/world expert in [area of expertise], I'd really appreciate it if you could check the accuracy of the [section name] section of this document, especially pages [page numbers].

Of course, you're welcome to look over the whole document, and comment on everything, including the grammar, but what I really need is for you to check the [section name] section.

To meet our shipping schedule, I'll need your corrections and comments by:

[date here, in nice big bold type].

Call or email me, and I'll come and pick it up.

Thanks in advance,

[your name here]

[contact info: phone number, extension, email address, etc.]

Other notes:

- If you can, deliver the document in person, introduce yourself and explain what you need from them.

- Keep a list of those you've given the document to.

- Send out a reminder, by email or in person, enough ahead of the due date so they can actually get it done.

- Make sure there are page numbers in the document.

- Put the reviewer's name on the document itself, in case the cover letter is lost or tossed.

- You may want to put the distribution date on the cover letter, but it might be best to just have the due date, so there's no confusion.

Appendices

Appendix 1: Recommended Reading

Writing and Editing

The Elements of Style, by Strunk and White
The Writer's Journey, by Christopher Vogler
On Writing Well, by William Zinsser
The Elements of Editing, by Arthur Plotnik
Pinkert's Practical Grammar, by Robert Pinckert
Technical Editing, by Judith A. Tarutz
Zen and the Art of Writing: Essays on Creativity, by Ray Bradbury
Comedy Writing Step by Step, by Gene Perret
Comedy Writing Secrets, by Melvin Helitzer
The Copywriter's Handbook, by Robert W. Bly
Writing for Children & Teenagers, by Lee Wyndham
The Craft of the Screenwriter, by John Brady
The Complete Book of Scriptwriting, by J. Michael Straczynski

Necessary Reference Books

Dictionary, Webster's 10th or later
Children's Dictionary
Thesaurus
Books of Quotes
Rhyming Dictionaries
Dictionaries of Slang and Euphemisms
Collections of Cliches
Almanacs
Joke Books
Comic Collections
Fowler's Modern English Usage
Wired Style, edited by Constance Hale
The Chicago Manual of Style
The New Well-Tempered Sentence, by Karen Elizabeth Gordon

The Transitive Vampire, by Karen Elizabeth Gordon
Children's Writer's Word Book, by Alijandra Mogilner

Interface Design (and design in general)

Macintosh Human Interface Guidelines, by Apple Computer, Inc.
The Windows Interface, Microsoft Press
TOG on Interface, by Bruce Tognazzini
The Elements of Friendly Software Design, by Paul Heckel
The Art of Human-Computer Interface Design, edited by Brenda Laurel
Computers as Theatre, by Brenda Laurel
Understanding Comics, by Scott McCloud
The Design (or Psychology) of Everyday Things, by Donald A. Norman
Turn Signals are the Facial Expressions of Automobiles, by Donald A. Norman
Things That Make Us Smart, by Donald A. Norman
The Visual Display of Quantitative Information, by Edward R. Tufte
Envisioning Information, by Edward R. Tufte
Anything by R. Buckminster Fuller (not easy reading)

The World, Technology and Our Times

The Popcorn Report, by Faith Popcorn
Future Shock, by Alvin Toffler
The Third Wave, by Alvin and Heidi Toffler
Mirror Worlds, by David Gelernter
Out of Control, by Kevin Kelly
The Dictionary of Cultural Literacy, by Hirsch, Kett, Trefil

Creativity

A Whack on the Side of the Head, by Roger von Oech
A Kick in the Seat of the Pants, by Roger von Oech
The Care and Feeding of Ideas, by Bill Backer
Drawing on the Right Side of the Brain, by Betty Edwards
Anything by Philip K. Dick or Philip Jose Farmer
The Oz books by L. Frank Baum are also marvels of creativity

Children's Books and Magazines

Anything by James Howe, L. Frank Baum, Bruce Coville and Roald Dahl. R. L.
   Stein, too.
Cricket (Magazine)
Nickelodeon Magazine

Popular Science

Chaos, by James Gleick

Three Scientists and Their Gods, by Robert Wright

The Blind Watchmaker, and other books by Richard Dawkins

The Cartoon Guide to Genetics, by Larry Gonick and Mark Wheelis

Calculus Gems: Brief Lives and Memorable Mathematics, by George F. Simmons.

The Case for Mars, by Robert Zubrin

Godel Escher Bach : An Eternal Golden Braid, by Douglas Hofstadter

The Minds I, by Douglas Hofstadter

A Brief History of Time, by Stephen Hawking

Genius, by James Gleick

Longitude, by Dava Sobel

Complexity, by Roger Lenin

Life of the Cosmos, by Rick Smolin

On Human Nature, by Edward O. Wilson

Broca's Brain, by Carl Sagan

How Things Work, and other books by by David MacAulay

The Naked Ape, by Desmond Morris

Powers of Ten, by Charles and Ray Eames

'Surely You're Joking, Mr. Feynman!' : Adventures of a Curious Character, by Edward Hutchings (Editor), Ralph Leighton, Richard Phillips Feynman, Albert Hibbs

Coming of Age in the Milky Way, by Timothy Ferris

The Man Who Mistook His Wife for a Hat : And Other Clinical Tales, by Oliver W. Sacks

Connections, and other books by James Burke.

Product Development

Debugging the Development Process, by Steve Maguire

Managing Software Maniacs, by Ken Whitaker

Rapid Development, by Steve McConnell

Dynamics of Software Development, by Jim McCarthy

Manuals

Maxis: *SimAnt, SimLife, El-Fish, Widget Workshop* and *SimCity 2000* are my favorites.

Apple: Most of the Macintosh manuals are pretty well done

Adobe: for wonderful examples of manuals that explain very complex subjects in a clear, well-organized way, read the manuals for *Photoshop* or *After Effects.*

Other Recommended Reading

You Just Don't Understand, by Deborah Tannen (about male/female language differences)

The Shockwave Rider, by John Brunner

Ender's Game, by Orson Scott Card

Something by William Gibson (short stories: *Burning Chrome*, 1st novel: *Neuromancer*)

Educational frameworks from various states, at least California and Texas

Read screenplays (not novelizations) of movies you've seen. (A lot of these are now available in stores. And check out www.script-o-rama.com)

Appendix 2: Contact Information

If you have writing methods or techniques that work for you or have a morsel of philosophy that other readers of this book would find useful, send them to me.

If you find a typo or other mistake, let me know. But please be nice.

If you send a message, please state the book title, edition and page number, if applicable.

You can contact me about this book in the following ways:

Email: Michael@untechnicalpress.com

US mail: Michael Bremer c/o UnTechnical Press
P.O. Box 272896
Concord, CA 94527

FAX: 925 825-4601

And check out the website: www.untechnicalpress.com

Appendix 3: Author Bio

Michael Bremer has written or edited the manuals, screen text, interactive help, teacher's guides, packaging and marketing materials for dozens of computer games, beginning with SimCity in 1989. For many years, he was Director of Creative Services at Maxis, Inc., heading up the Writing, Audio/Music/Video and Internationalization groups. After a stint as a Senior Game Designer at Electronic Arts, he is now an independent writer and publisher, and a founding member of the Remedial Film School.

Index

Order Form

Fax orders—1 925 825-4601

Online orders—
www.untechnicalpress.com

Telephone orders—Call toll free:
1 888 59 BOOKS (592-6657)
Have your VISA, MasterCard, or
AMEX ready.

Postal orders—
UnTechnical Press,
P.O. Box 272896,
Concord, CA 94527, USA
Telephone: 925 825-4601

Please send the following books. I know that I may return any books for a full refund.

See our complete line of books at www. untechnicalpress.com.

Quant.	Title	Unit Price	Total Price
	UnTechnical Writing—How to Write About Technical Subjects and Products So Anyone Can Understand	$14.95	$.
	The User Manual Manual—How to Research, Write, Test, Edit and Produce a Software Manual	$29.95	$.
			$.
			$.
			$.
		Subtotal	$.
		*Sales Tax	$.
		**Shipping	$.
		Total	$.

*Sales Tax: Add 8.25% sales tax for books shipped to California addresses.

**Shipping: For books shipped to locations inside the United States, please include $4.00 for the first book and $2.00 for each additional book. Call for shipping charges for locations outside the United States.

Payment:

❑ Check enclosed, payable to **UnTechnical Press** (Please write phone number and driver's license number on the check to avoid a shipping delay.)

❑ Credit Card: ❑ VISA   ❑ MasterCard   ❑ AMEX

Card number: _____

Name on card: _____ Exp. Date: _____ / _____

Cardholder's signature: _____

Ship to: Name: _____

Address: _____

_____

City: _____ State: _____ Zip: _____